Mihail Ispan

Tellus Plana
Territory of Freedom

Table of Contents

About the author..7

I. Introduction..9

II. Historical Summary..11

III: Etymology..17

IV. Definitions..21

V. Succinct Explanations..25

VI. The convex-heliocentric theory...................................33

VII. The Visual Matrix..41

VIII. Tellus Plana..63

IX. The Sun..93

X. The Moon..107

XI. The Black Sun..143

XII. Mysteries of the Stars..145

XIII. Questions and Answers..157

Imprint...173

About the author

I was born and studied in Romania. After my studies I obtained my engineering degree. After university I worked for several years as an engineer in Romania and then settled in Germany. Here I attended for several years specialisation courses in computer science. I have always enjoyed studying and experimenting with new things autodidactically. That is why I have always had several extra-curricular interests that I have pursued with passion, including sports, painting and the study of various scientific disciplines. I have been practising spirituality for more than twenty years and for some years I have also been writing. The works I have written arise from a holistic way of thinking and from my unconditional, immeasurable and eternal love for all human beings.

I. Introduction

In this paper I present information about the Earth we live on. This information is quite astonishing to the average man because it doesn't agree with what you learn in school, high school or college, nor with astronomy books or what you hear, read or see in the media. This information will surprise even the most educated, informed and open-minded people. That is why it will cause many readers to experience the phenomenon of cognitive dissonance, which consists in their firm rejection and the impression that they have been produced by the fantasy of a man with a boundless imagination. However, all this information is in line with reality. I therefore recommend that you do not allow yourself to be misled by the feeling that they are pure fantasy, but that you read them carefully to the end because all this information is scientific in nature, is consistent with truth and reality, has logical continuity, you can perceive it yourself, you can prove it experimentally yourself, it is supported by unquestionable evidence and it makes sense. My recommendation is to read this work very carefully at least twice. After the first reading I recommend that you investigate for yourself in the real world both by careful observation and perception of the phenomena explained in this work and possibly by experimentation. After that, it is good to read this work again to understand the phenomena in depth and to be able to make the necessary connections correctly. This book is currently the only book on this planet that provides a model of the Flat Earth that matches reality and in which all explanations are expressed in a scientific, coherent and understandable way. In addition, in this book are the experimental methods by which you can convince yourself that my statements are correct.

Note: In this work, the images are not made to scale because they are representations of very large dimensions, usually they are representations of terrestrial or cosmic dimensions in relation to small dimensions, for example relief or buildings. Because these images represent very large dimensions in relation to small dimensions it is very difficult, in most cases impossible, to represent them to scale. The images in this paper provide additional information to understand the ideas they refer to.

II. Historical Summary

On our planet, humanity has for a long time been held in slavery by a handful of people calling themselves emperors, kings, queens, pharaohs, presidents of states and authorities. The authorities have always used violence, terror and fear to keep people in slavery. To escape from this slavery, the people had only three options: revolt, escape from the master's territory and death. There was no other alternative. There were people who chose the first option. With the weapons they had at hand, they tried to free themselves from the ruthless slavery that was destroying their lives, their families and their entire community. Most of the time, however, they were defeated and had to suffer the consequences of their rebellion, which were torture, slavery, imprisonment and death. They were defeated because the authorities had many well-armed slaves on their side who they put to fight against the rebellious slaves. Other slaves, fed up with the miserable life of slavery, chose the second option: the road of wandering. Some were caught and sent back into slavery, others died on the way, but not a few found hospitable and unpopulated territories where they could settle, build a new community and live in freedom. The third category of slaves was made up of slaves who were so bitter about their situation that they chose death by suicide.

The authorities knew that some slaves were leaving their estates and realised that slavery had to be not only physical but also mental. In order to defeat the will of the slaves to flee from their estates, the authorities told the people that the horizon line is the edge of the world and whoever ventures into the unknown and crosses this edge will fall into the endless abyss of the cosmos and be lost forever. Believing this, people were afraid to venture into the unknown, they gave up the thought of leaving and accepted their fate of remaining poor and humble slaves forever. During this time, the authorities sent well-equipped ships out to sea to search out distant territories in the hope of finding new territories, wealth and slaves. The lie of the authorities was believed for a very long time, for hundreds of years, but there came a time when people realised that the horizon line is not the edge of the world because the higher you rise, the more territory you see. Consequently, if the observer's eyes go higher and higher, the horizon line recedes and therefore it cannot be the edge of the world as the authorities preached. In the meantime, the authorities have developed their sea navigation. People did not have the courage to build large boats to sail out to sea because they feared they would fall into the endless abyss of the cosmos.

In this way, the authorities of the world began to work together to develop a common force and thus be better able to exercise their authority to keep people enslaved. But the authorities had a problem! People no longer believed that the horizon was the edge of the world and tended to create plans to wander off in the hope of finding a better place under the Sun. So the authorities had to create another theory that would destroy the slaves' hope of finding new territories. This gave rise to the astronomical theory of the spherical Earth, which says that the Earth is a sphere, that it floats in outer space, and that it is so far away from the other planets that it is impossible for ordinary people to get through. This theory was introduced in school so that, as young children, people believed it and never had any hope of ever escaping from slavery. Lately, more and more people understand that the authorities' astronomical theory is wrong and that the Earth is flat. That is why the authorities are currently playing both ends. Some people of the authorities are still trying to maintain the astronomical theory of a spherical Earth, other people of the same authorities admit that the Earth is flat but go on to say that it would be covered by a dome suggesting to people that there would be no hope of escaping their clutches. Fortunately for humans, the Creator of the Universe created the Earth in such a way that humans can never be made slaves and when this is attempted, they may leave for other lands to live in freedom. The Earth is neither spherical nor bounded. The earth is flat, immobile and infinite. On the Flat, Immobile and Infinite Earth there are a huge number of territories similar to our planet.

Beyond Antarctica lie new planets. An increasing number of documents are appearing lately, such as photos and videos taken in China and North America showing two suns instead of just one. One

II. Historical Summary

of them is our Sun, while the other is the Sun belonging to a neighbouring planet. This is not the optical illusion where three lights appear: one in the middle which is the sun and two on the sides which appear to be half suns. The two suns that appeared in China and North America are two distinct and real suns. The world's authorities are well aware that the Earth is flat. You can figure this out by looking at their flags, emblems and logos on which the Earth is depicted as flat. These logos are: the emblem of the United Nations (UN), of the International Civil Aviation Organization (ICAO), of the International Maritime Organization (IMO), of the World Meteorological Organization (WMO) and of the World Health Organization (WHO).

Knowing that the Earth is flat, authorities have always suspected that new territories lie beyond the high ice walls of Antarctica. Numerous expeditions have been organised to reach the southern polar region, to set foot on Antarctica and to build colonies there. Between 1946 and 1947, the US-sponsored Operation Highjump (OpHjp) was the largest expedition to Antarctica. This operation began in August 1946, was organised by the Freemason Admiral Richard E. Byrd and was co-financed by the Rockefeller house. An armada of 5,000 men, 13 naval vessels and 26 aircraft participated in the operation. On his return home Admiral Richard E. Byrd gave television interviews in which he revealed that beyond the South Pole lay a continent larger than North America. Because he said more than he should have, he was hospitalized and killed at the behest of his older Masonic brothers. The activities in Antarctica are governed by the Antarctic Treaty and some 200 additional agreements signed by several countries around the world. The articles of the Antarctic Treaty set out excellent strategic objectives, such as that Antarctica should always be used for peaceful purposes and for the benefit of all mankind. From a vast historical experience, everyone knows that everything that the authorities put down in writing, whether treaties or laws, is dust in the eyes of the people.

There are many indications suggesting that the exploration of Antarctica is severely constrained by the need for a permit and by the military personnel stationed in Antarctica. Information, which cannot be confirmed from independent and competent sources, says that tourism on one's own is practically forbidden in Antarctica, that private and unauthorised flights over Antarctica are prohibited, and that independent researchers who have attempted to explore Antarctica, such as Norwegian Jarle Andhøy (b. 23.10.1977) and his crew, have been threatened with violence, arrested or even killed. According to these indications, the authorities are preparing an encirclement of humanity by placing military bases in Antarctica in order to prevent people from migrating to the new territories.

The worst situation for the authorities is when enslaved people leave the territory they control and migrate to new territories because, in this case, those who form the authorities will have to work themselves in order to support themselves, which would be a disastrous situation for them considering that they have never worked. The world's super-rich have already built residences in the new territories beyond Antarctica.

II. Historical Summary

RSRM rocket   Rocket launch

They travel to these territories by rocket, plane and land transport. The manned RSRM (Reusable Solid Rocket Motor and Solid Rocket Boosters) rockets built by NASA consist of an orbiter, which is actually an aircraft, and tubes containing solid fuel.

These rockets are reusable. They are launched and lift off to a very high altitude. After this the fuel tubes are disconnected and the so-called orbiter, which is actually an aircraft, flies with its own engines to new territories. This can be seen from the fact that the trajectory of these rockets becomes a curve pointing towards the ground. The „scientific" authorities say that the trajectory of the rockets looks like a curve because these rockets fly parallel to the curve of the earth. The curve of the Earth does not exist because the Earth is flat. The rockets fly on a curved trajectory to steer the plane towards new territories. Sometimes the world's super-rich want to travel to the new territories using ground vehicles. In addition to other purposes, Antarctic bases serve as intermediate stations for ground transport to new territories. Some people ask the following question: Why do the masters of the world destroy or allow the destruction of our planet? Why are the air, water and soil poisoned? Why are forests being destroyed? These things will affect them too!

The answer to this question is very simple. The masters of the world have residences in territories beyond Antarctica where there is excellent air, water and soil. They don't care what happens on this planet. The super-rich of the World use this planet as a surface and deep mine from which they extract with the help of humans and robots the raw materials they need. They also use it as a production site. Humans are their slaves.

III: Etymology

In our age, which can rightly be considered the age of lies, people are convinced that the Earth is a convex-shaped object and that it executes various motions, not because they can provide explanations based on their senses and observations in favour of this bizarre assumption, but because so-called science has indoctrinated them from an early age to believe that only its abstract and false terms would be real and nothing that is obvious to human perceptions would be true. In the following I will explain the etymology of the main terms used in this book.

Earth: In Romanian, the word Pământ (Earth) is related to the Latin term pavimentum which means: hard-packed and levelled ground, flagstone or mosaic floor, pavement, floor, paved road and smooth place. All these latter notions describe very accurately the land we live on. The convex-heliocentric theory starts from the false assumption that the Earth is convex in shape. The Earth does not have a convex shape, but is perfectly flat when you remove landforms such as mountains, volcanoes, hills, valleys, ripples on the surface of the water and undulating crust on the surface of the Earth.

Tellus: The Latin term Tellus means Earth. The etymology of this term is uncertain. It seems to be related to the Sanskrit term talam which means: flat land and plain. Interestingly, in ancient Roman religion and myths there was both the term Tellus Mater and Terra Mater. This could mean that ancient people made a distinction between Earth and Planet. It is possible that the Latin term Tellus Mater referred to the Earth as an infinite area on which a multitude of planets are found. The Latin term Terra Mater is known to refer to the planet on which we live. In other words, the term Tellus Mater could have meant Mother Earth, while the term Terra Mater could have meant Mother Planet.

Planet: The term planet is related to the Latin term planus which means flat, level, even, plain. This term indicates that the planet is flat.

Sferoid: The term spheroid comes from the merging of the term sphere with the suffix oid. The suffix oid has the following semification: in the shape of. The term spheroid means: in the shape of a sphere or a sphere-like shape.

Geoid: The term geoid comes from the merging of the term geo with the suffix oid. The term geo is related to the term gaia and means: land, territory and soil. It can be used as a prefix to form nouns such as geography, geology, geometry, etc. The suffix oid has the meaning: in the form of.

Oblat: The term oblate is an adjective. It is related to the Latin term oblatus. The Latin term oblatus is derived by joining the term ob with the term latus. The term ob means: from edge to edge, along and across. The term latus means wide. The term oblat like the term oblatus means: more wide than long.

Oblong: The term oblong is an adjective. It is related to the Latin term oblongus. The Latin term oblongus is derived by joining the term ob with the term long. The term ob means: from edge to edge, along and across. The term longus means long. The term oblong as oblongus means: longer than wide.

Sun: In many languages the word Sun begins with the letter S.

III: Etymology

If you photograph the Sun at the same time of day and every few days you will get an analemma. The analemma has the same shape as the infinity symbol in mathematics. It can be obtained by combining a letter S with another letter S rotated symmetrically to the previous one. The letter S in the term Sun literally symbolises the analemma formed by the Sun.

IV. Definitions

In this chapter I give you the definitions of the terms I will use in this book, including explanations. Since all the theories about Earth and outer space offered by so-called scientific authorities are false, many of the concepts used in these theories have wrong meanings. Consequently, I have to redefine many of these notions so that they correspond to reality. All the information from antiquity tells us that the Earth is flat and immobile. Ancient people had very advanced science and relied on the evidence delivered by their senses.

Earth: The term Earth refers to a flat, immobile, infinite area on which many planets are found.

Flat Earth: The phrase Flat Earth refers to the fact that the Earth's surface is flat if we ignore the relief of mountains, hills, plains, valleys, depressions, plateaus, etc., if we ignore the waves of the seas and oceans, and if we ignore the complex wavy crust of the Earth. We ignore all these because their size relative to the size of the Earth and our planet is extraordinarily small.

Buoyant force: Buoyant force is the force acting on a body immersed in a fluid. The vector of the buoyant force is vertical and the direction of this vector is from bottom to top. The buoyant force is equal to the weight of the volume of fluid displaced by the submerged body.

Telluric force: The convex-heliocentric theory assumes that the planet would have a gravitational force which would be a mutual attractive force of two entities whereby, in the case of supposed celestial bodies, one entity could gravitate around the other. Since there are no celestial bodies and celestial entities do not gravitate around other celestial entities, the terms gravitational force and gravity are inappropriate. There is no such gravitational force, but the Earth has an attractive force which, if it is greater than the buoyant force, causes bodies in the air to fall. The Earth's attractive force works perfectly well on a flat Earth, but it does not work on a convex planet. On a convex planet, lakes, seas and oceans are supposed to be domed which would cause them to collapse due to their weight and flood the earth. In reality, however, lakes, seas and oceans do not collapse under their own weight and therefore do not flood the ground. If you let a heavy body fall from a certain height, its initial velocity will be 0 m/s. After some time its speed will be greater than

0 m/s. As a result, the speed of this body has increased. The increase in velocity is actually acceleration. As you can see, this body will fall in an accelerated manner. Where there is acceleration, there must also be force. This force is the Earth's attractive force, which I will call the telluric force.

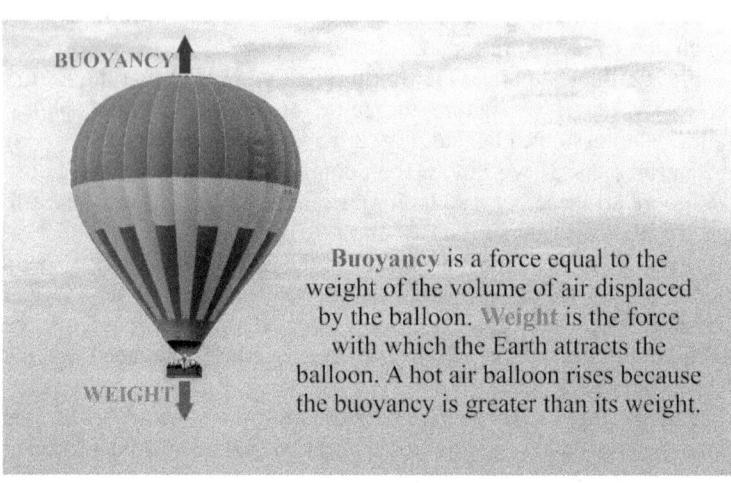

Buoyancy is a force equal to the weight of the volume of air displaced by the balloon. Weight is the force with which the Earth attracts the balloon. A hot air balloon rises because the buoyancy is greater than its weight.

Planet: As stated in the Etymology chapter, the term planet is related to the Latin word planus, which means flat, level, smooth, plain. Consequently, this term also states that the planet is basically flat and therefore not convex. The term planet means a territory on the flat, immobile and infinite Earth. There is a complex crust on the planet. Above this crust is the relief of the planet. The topography is made up of mountains, hills, plains, valleys, depressions, plateaus, etc. Water in the form of flowing and standing water is found on the planet's topography. The planet's topography is home to a variety of life. The term planet refers to a finite territory on Earth. This territory is roughly in the shape of a circle. A celestial entity called the Sun

IV. Definitions

revolves above the planet. On Earth, the number of planets is constantly increasing and decreasing.

Globist: The term globist refers to a believer in the convex-heliocentric theory. In this book, the term globist has no pejorative meaning.

Globalist: The term globalist refers to someone who belongs to both the convex-heliocentric theory community and the globalist force of powerful people on the planet who have a common agenda and are trying to implement a political, economic, technological and social dictatorship on this planet. Both the proponents of the convex-heliocentric theory and those who want a global dictatorship are in fact the same people.

Platist: The term platist refers to someone who believes in theories that claim the earth is flat. In this book, the term platist has no pejorative meaning.

Axial precession: The convex-heliocentric theory assumes that the planet, in addition to other motions, would have an axial precession motion. The axial precession motion would consist of the planetary axis rotating around another axis. A single cycle of this precessional motion would last about 25 772 years.

Orbital-tangential speed: Orbital-tangential speed is the speed of an entity moving around a point at a distance from that entity. This speed is the distance, expressed in a unit of length, per unit of time.

Orbital-angular speed: Orbital-angular speed is the number of radians, executed by a body, per unit time.

V. Succinct Explanations

In this chapter I give you a brief explanation relating the flat, immobile and infinite earth.

Flat, Immovable and Infinite Earth: The real model of the Earth I will call the flat, immobile, infinite Earth model.

Lack of planetary convexity: The upper surface of the water is flat and not convex. Rational people perceive the upper surface of Lake Razelm to be flat, but the convex-heliocentric theory argues that we should not trust our eyes because this lake, like other lakes and seas and oceans on the planet, is convex because it is on a convex planet. Mirabile dictu!

Large length constructions: All canals, tunnels and railways are built without taking into account the supposed convexity or curvature of the planet.

Lighthouses: The light from lighthouses can be seen at very great distances. From these distances it would not be visible if the planet were shaped like a convex object because of the curvature of the planet.

North Pole: The shape of the planet, seen from above, is similar to that of a circle. On the flat planet, the North Pole is the magnetic centre of the planet and lies at its geometric centre.

South Pole: The South Pole does not exist on the flat planet as a geographical point. The so-called South Pole is in fact the giant wall of ice of Antarctica that lies all around the planet.

Convex-heliocentric theory: All the elements of the scholastic astronomical theory are foggy. In this theory even the shape of the planet we live on is not clearly explained. NASA suggests through its images that its shape is perfectly spherical. Some astronomers say it is shaped like an oblate spheroid, others say it is shaped like a geoid. One so-called astrophysicist says it is pear-shaped. The sphere, oblate spheroid, geoid and pear are convex bodies. convex-heliocentric astronomers also claim that this planet, in addition to other rotational motions, would rotate around the Sun. These are the reasons why I will call the scholastic astronomical theory: the convex-heliocentric theory.

Orbital-tangential and orbital-angular velocity: The image on the left shows the orbital rotational motion of an entity as seen from above. The image on the right shows the orbital rotational motion of an entity, seen from the side. The path of an entity moving around a point at a distance from that entity is called an orbit. Speed is the ratio of the distance travelled by an entity to the time in which the entity travels that distance.

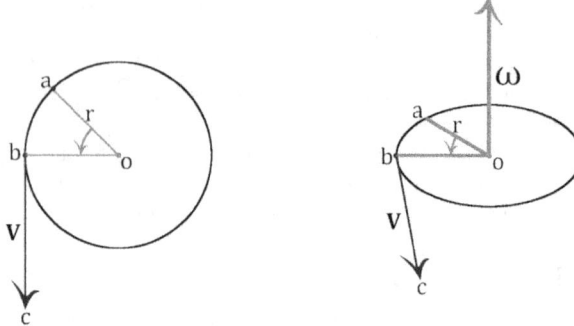

Orbital-tangential velocity and orbital-angular velocity are the velocities of an entity moving around a point at a distance from that entity. Orbital-tangential velocity is the distance expressed in a unit length per unit time. Orbital-angular velocity is the number of radians per unit time. The elements in these images are as follows:

.the plane formed by the trajectory of the entity is called the plane of rotation.
.o is the point around which that entity moves. It is called the rotation point.
.a is the point through which the entity passes at time t_1.
.b is the point through which the entity passes at time t_2.
.α is the angle between line oa and ob. The arc of this angle is traversed by the entity in the time interval $t_2 - t_1$. This time interval can be denoted by t.

V. Succinct Explanations

.**V** is the orbital-tangential velocity vector.
.ω is the vector of the orbital-angular velocity.
.c is the tip of the vector **V**.
.d is the tip of the vector ω.
.r is the radius of rotation.
.the direction of the vector **V** is the line through bc. This direction is always tangent to the orbit and is always changing.
.the direction of the vector ω is always in the centre of the plane of rotation and is always perpendicular to this plane. In this case, the direction of this vector is the straight line passing through od.
.the distance bc is the modulus of the vector V. The modulus of the vector V is the scalar magnitude of this vector. This is expressed, for example, as 7.9 km/s.
.the distance od is the modulus of the vector ω. The modulus of the vector ω is the scalar magnitude of this vector. This is expressed, for example, as 15 degrees/hour or 2¶/24 = 0.26 radians/hour.

.the arrow inside the angle α indicates the direction of rotation in which the entity is moving.
.the sense of the vectors is indicated by corresponding arrow.
.the sense of the vector **V** is always changing, but in such a way that the vector ω always has the same sense.
.the sense of the vector ω is always the sense given by the linear movement of a corkscrew relative to the plug it enters or exits. Example: if you put a corkscrew through the stopper by turning the corkscrew clockwise, then the corkscrew will enter the stopper which means that the direction of the vector ω will be from you towards the stopper.

Angle can be expressed in degrees and radians. Expressed in degrees, a circle is 360 degrees. Expressed in radians, a circle is 2¶. Therefore, 360 degrees = 2¶. 1 radian = 360/2¶. The decimal representation of ¶ truncated to 4 decimal places is: 3.1415. If the angle of rotation is expressed in radians, then the following mathematical relationship exists between orbital-tangential and orbital-angular velocity: $\mathbf{V} = \mathbf{\omega} \times \mathbf{r}$.

The Sun: The Sun moves on an imaginary cylindrical path, clockwise, up and down over the flat planet on the surface of an imaginary cylinder with the bottom base above the Tropic of Capricorn and the top base above the Tropic of Cancer. In this case, the orbital-tangential and orbital-angular velocities remain unchanged at all times. This motion is shown in the image above. The perspective causes the Sun's circular paths to decrease with increasing altitude.

Because of this, the observer sees the imaginary cylinder as a cone. In this case, the Sun spirals clockwise up and down above the flat planet on the surface of an imaginary truncated cone with the high base above the Tropic of Capricorn and the high base above the Tropic of Cancer. The radius of the orbit in which the Sun moves shrinks as the Sun's altitude increases. Since the radius of the orbit changes with the altitude of the Sun, the Sun's orbital-tangential velocity also changes, but the Sun's orbital-angular velocity remains constant.

V. Succinct Explanations

As the Sun ascends the truncated cone, the Sun's orbital-tangential velocity appears to become increasingly slower, and as the Sun descends the truncated cone, the Sun's orbital-tangential velocity appears to become increasingly faster. From this graph we can see that the perspective changes the Sun's orbital-tangential velocity. But the change in velocity is only an appearance produced by perspective. In reality, the Sun's orbital-tangential velocity always remains the same. Circular motion on the horizontal produces the day-night cycle, while vertical to-and-fro motion produces the seasons.

Moon: The Moon is the Sun's reflection on the Earth's atmosphere. The motion of the Moon is similar to the motion of the Sun.

Sun and Moon: The convex-heliocentric theory assumes that the Sun would be at a distance of 149,600,000 km and that it would illuminate the Moon. This would be why we would see the Moon. If this were the case, the Moon would be seen as a small bright spot and not as a circle during a Full Moon. The Sun's rays would illuminate the spherical half of the Moon, but only a small spot of light would reach the observer's eye because, due to the Moon's convexity, the Moon's other reflections would be scattered in directions other than towards the observer's eye. Consequently, the Moon should be, according to convex-heliocentric theory, a small spot of light in the sky.

Stars: The stars revolve above a flat, immobile planet.

Black Sun: The Black Sun is a black disc that produces the solar and lunar eclipses.

Convex-heliocentric theory: The convex-heliocentric theory is a pseudoscientific theory which assumes that the planet is shaped like a sphere, oblate spheroid, geoid or pear.

Oblate spheroid: As stated in the Etymology chapter, the term spheroid comes from the merging of the term sphere with the suffix oid. The suffix oid has the following meaning: in the form of. The term spheroid means: a regular and sphere-like shape. The spheroid can also be called an ellipsoid of revolution. The spheroid may be oblate or oblong. The oblate spheroid has the shape of a sphere flattened at the poles. Some astronomers say the planet is shaped like an oblate spheroid.

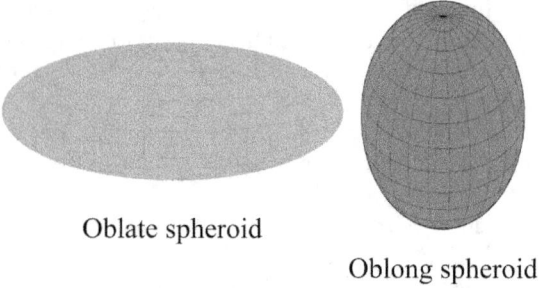

Oblate spheroid

Oblong spheroid

Geoid: The geoid is a three-dimensional, irregular shape. According to the convex-heliocentric theory, the geoid would have the shape that the water on the planet would take under the influence of gravity and the rotational motions of the planet without taking into account the influence of winds and tides. Some astronomers argue that the planet is shaped like a geoid.

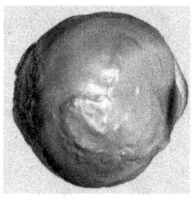

South Pole: A South Pole at the lowest point of a supposedly convex planet does not exist.

V. Succinct Explanations

Anomalies between the Arctic and Antarctic: Various anomalies and differences between the Arctic and Antarctica, such as the Midnight Sun in the Arctic, prove that the planet cannot be convex.

Moon and Mars landing: Man has not been to the Moon. Moon and Mars landing were hoaxes staged by Freemasons. The footage of these hoaxes was taken on this planet. The supposed artificial satellites placed in our planet's so-called orbit and the supposed space stations do not exist. All the films and images by the NASA space agency showing: the Hubble Space Telescope, the ISS space station, planets and so on are in fact computer-generated images. The so-called astronauts are just actors. Some images of these so-called astronauts floating in non-gravitational space are created using cables with actors attached to them. Other images are created in planes called Zero-Gs flying on a parabolic curve to produce apparent non-gravity. Other images are produced in a large pool of water in Texas called the Neutral Buoyancy Laboratory, where non-gravity is simulated.

Experimental evidence: There is as yet no experimental evidence that the planet is convex and spinning through space, but there is plenty of experimental evidence that the planet is flat and immobile. Modern astronomy has never provided any evidence for the accuracy of its theories. It presents many mathematically correct calculations that start from false premises. Because the premises are false, their calculations lead to false results.

International astronomical community: The international astronomical community and the world's space agencies are controlled by Freemasons. The space agency NASA is the headquarters of the world's space agencies. The Hebrew term nasa means: to deceive in a big way. On the emblem of this space agency are stars and a serpent's tongue. The translation of the symbolism of this emblem is as follows: to deceive people in a big way about the Universe, to deceive people in a big way about astronomy.

VI. The convex-heliocentric theory

In this chapter I will set out the assumptions of the scholastic type of astronomical theory, which can also be called the convex-heliocentric theory. Also here I will briefly explain the reasons why this theory is only a theory and not scientific.

Planetary Convexity: The convex-heliocentric theory holds that the planet would have the shape of a convex object with an average circumference of 40,041.47 km which means that the height that would be occulted by the curvature would be 8 cm for the first kilometre, 32 cm for the second kilometre, 72 cm for the third kilometre and so on according to the rule: square the number of kilometres between the observer and the observed target then multiply the result by 8. The convex-heliocentric theory is not in line with reality because this supposed occultation of distant objects is totally erroneous both on land and on water and is therefore not taken into account in the construction of canals, tunnels and railways, nor in nautical navigation nor in aeronautical navigation.

Planetary motion: The convex-heliocentric theory holds that the planet would perform an enormously complex motion consisting of 6 different motions. **The first** motion would be the rotation of the planet around its axis so that a point on the planet's surface at the equator would have an average speed of 1,675 km/h (info: speed of sound = 1,224 km/h). **The second** motion would be axial precession motion, which would consist of the planet's axis spinning around another axis, making the planet spin like a peg top that has lost some of its speed. A complete rotation of the axial precession motion would take about 25,772 years. **The third** motion would be the revolution of the planet, which would consist of a rotational motion of the planet around the Sun at an average orbital speed of 110,000 km/h. **The fourth** motion would be the motion of the planet, which would occur as the entire Solar System rotates around the galactic centre. This motion would be at a speed of 828 000 km/h. **The fifth** motion would be the motion of the planet, which would take place in conjunction with the rotational motion of our galaxy, the Milky Way, around the galactic axis. This motion would be performed at an average orbital speed of 220 km/h. **The sixth** motion would be the motion of the planet in conjunction with the movement of the Milky

Way galaxy towards the constellation Hydra. This motion would be at a speed of 2.15 million km/h.

Sun: In the convex-heliocentric theory, the Sun would be a plasma sphere at a distance of 149,600,000 km.

Moon: In the convex-heliocentric theory, the Moon would be a solid body orbiting the planet, and so it would be a natural satellite of the planet.

Eclipses: In the convex-heliocentric theory, eclipses are explained as follows. The solar eclipse would be due to the Moon passing in front of the Sun. The lunar eclipse would be the result of the Sun, planet and Moon being aligned in a single line so that the planet would be in the middle and would block the Moon. The explanation of eclipses given by the convex-heliocentric theory is not correct. The eclipses of the Sun and Moon are due to a celestial entity. Globist astronomers, specifically those at NASA, pride themselves on the fact that they can predict eclipses, and people believe that this prediction would be based on the parameters of the convex-heliocentric theory which leads to the conclusion that this theory would be the correct one. The truth is that NASA astronomers use ancient models to calculate eclipses and that there are no specific parameters of convex-heliocentric theory in these models. Consequently, the prediction of eclipses has nothing to do with the convex-heliocentric theory and therefore it is not evidence that this theory is correct.

Earth's attractive force: The convex-heliocentric theory assumes that the planet would have a gravitational force which would be a mutual attractive force of two entities whereby, in the case of supposed celestial bodies, one entity could gravitate around the other. Since there are no celestial bodies and celestial entities do not gravitate around other celestial entities, the terms gravitational force and gravity are inappropriate. There is no such gravitational force, but the Earth has an attractive force which, if it is greater than the buoyant force, causes bodies in the air to fall to the ground. The

VI. The convex-heliocentric theory

Earth's attractive force works perfectly on a flat Earth, but it does not work on a convex planet. On a convex planet, lakes, seas and oceans are supposed to be vaulted, causing them to collapse due to their weight and flood the land. In reality, lakes, seas and oceans do not collapse under their own weight and so do not flood the ground. If you let a heavy body fall from a certain height, its initial speed will be 0 m/s. After some time its speed will be greater than 0 m/s. Consequently, the speed of this body has increased. The increase in speed is actually acceleration. As you can see, this body will fall in an accelerated manner.

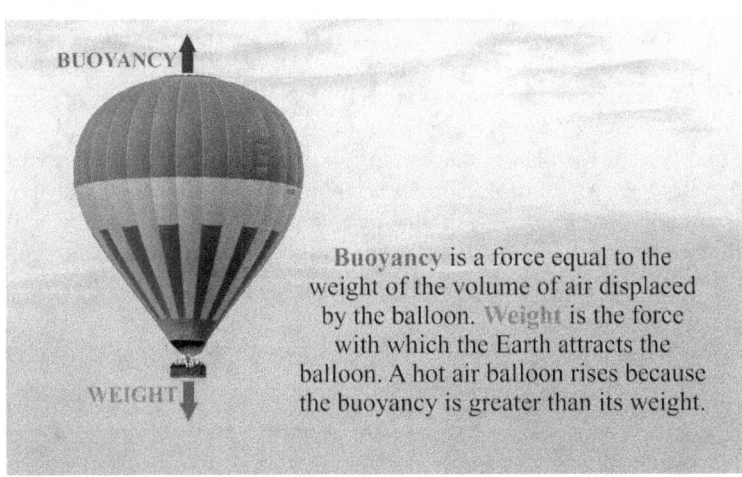

Buoyancy is a force equal to the weight of the volume of air displaced by the balloon. Weight is the force with which the Earth attracts the balloon. A hot air balloon rises because the buoyancy is greater than its weight.

Where there is acceleration, there must also be force. This force is the Earth's attractive force, which can be called the telluric force.

Sun and Moon: The convex-heliocentric theory assumes that the Sun would be at a distance of 149,600,000 km and that it would illuminate the Moon. This would be why we would see the Moon. If this were the case, the Moon would be seen as a small bright spot and not as a bright circle as we see it when the Full Moon is out. The

Sun's rays would illuminate the spherical half of the Moon, but only a small spot of light would reach the observer's eye because, due to the Moon's convexity, the Moon's other reflections would be scattered in directions other than towards the observer's eye. Accordingly, the Moon should be, according to convex-heliocentric theory, a small spot of light in the sky. In the following.

   I will elaborate on this phenomenon by example and by giving you an experiment you can do to convince yourself that, according to the convex-heliocentric theory, the Moon should be a small spot in the sky and not as it is seen in reality when it is a Full Moon. The following picture shows a section through a room. In this room there is an observer, a light source and a sphere. The centre of the sphere is at point O. The light source illuminates the whole room, including the sphere. From point S and around this point a small area of light will come towards the observer's eyes in the form of a small spot of light which will have maximum brightness. The observer will therefore see a spot of maximum brightness at point S. This light spot is actually the reflection of the light source. The light source will illuminate the floor, side walls and ceiling of the room. The light from the floor, side walls and ceiling of the room will also illuminate the sphere.

VI. The convex-heliocentric theory

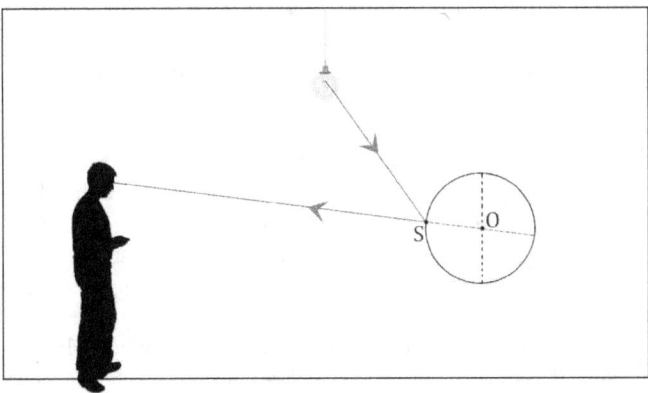

The sphere will reflect this light as a larger area of light but of lower brightness than the brightness of the small light spot. This light, coming from the floor, walls and ceiling, will surround the small light spot on the sphere. In this way, the observer will see the half sphere in front of him. In the picture below you can see how the observer sees the half-sphere in front of him.

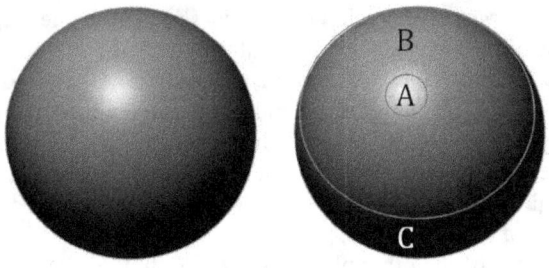

Circle A contains the maximum brightness spot coming from the light source. In the ellipse B is framed the area of light coming from the floor, the side walls and the ceiling of the room, and in the dark lunula C is framed the area from which no light is coming towards the observer's eyes. According to the convex-heliocentric theory, the Moon would float in outer space and be illuminated by the Sun. In a room there are surfaces, namely the floor, the side walls and the ceiling of the room. These surfaces reflect the light coming from the light source, and the light coming from these surfaces illuminates the sphere producing ellipse B. There are no such surfaces in outer space, which means that the elliptical surface of light is completely missing from the surface of the illuminated object. Therefore, according to the convex-heliocentric theory, the Moon should look like a spot of light surrounded by a black surface. Because there are so many atmospheric layers between the observer and the Moon, the Moon's black surface cannot be seen.

To better understand this I will use an example from painting. Imagine painting a blue sky on a canvas! After you finish painting the sky, you paint a black circle that would represent an unlit body in the sky. To simulate the atmospheric layers that exist between you and the unlit body, you would brush a blue layer over the black circle. After this layer dries, you apply a new blue layer in the same way and so on. At some point the black circle will disappear under the blue layers and all you will see is sky. The same happens in reality. Because there are so many atmospheric layers between you and the dark body, you won't be able to see the dark body. Remember this important point: only bright and illuminated entities can be seen in the sky, whether it is day or night. Consequently, it follows from the convex-heliocentric theory that the Moon should appear as a small bright spot in the sky. The Moon appears in the sky as a bright circle when it is a full Moon.

In the case of the other phases of the Moon, the Moon appears as a luminous lunula or as a quarter, Waxing Crescent or Waning

VI. The convex-heliocentric theory

Crescent. The Moon never appears as a small bright spot. Therefore, the assumption of the convex-heliocentric theory that the Moon is a solid entity floating in space and illuminated by the Sun cannot be correct. The Moon is a light. It is the projection of the Sun onto the Earth's atmosphere.

VII. The Visual Matrix

The Visual Matrix, which can also be called the structure of Perspective, is the structure of the image from the external reality of the human body. This structure is produced by the visual system. The mind interprets perspective according to the information it holds. If the mind has the wrong information, perspective will be misinterpreted. If the mind has incomplete information, the perspective will be interpreted incompletely. In our society, all people are indoctrinated from childhood and so there is a lot of erroneous and incomplete information in their minds. Consequently, the perspective is misinterpreted or incompletely interpreted by the vast majority of people. This is why most people do not understand that the Earth is flat, immobile and infinite.

The interpretation of the observed phenomenon comes from the mind. Understanding this phenomenon comes from the spirit. If you want to understand a phenomenon in depth, it is good to observe it for a longer time without judging. Meditate, to stop the mental-emotional flow of the mind loaded with a lot of erroneous information that clouds understanding. After some time, deep understanding comes effortlessly, from the spirit and by itself. Perspective is a visual matrix that produces a very powerful illusion.

This illusion is so powerful that so far most people do not understand it at all, and the rest only partially understand it. This book is the only document in the history of mankind in which correct information about the Perspective, Sun, Moon and Black Sun is presented. It will be decades, or perhaps even centuries, before this information is understood by the vast majority of people. But back to perspective.

We have therefore said that perspective is the structure of the image coming from the external reality of the human body and that this structure is produced by the visual system. Consequently, it is the visual system that produces perspective and so to understand perspective it is good to understand something about the visual system. When an observer looks at the sky, he is looking at a cosmic panorama that has supergiant dimensions.

This panorama is projected onto the retina of his eye which is about 22 mm in diameter. Because the cosmic panorama is supergiant in size and the human eye has a retina with a diameter of only about 22 mm, the cosmic panorama cannot be projected exactly as it is on the retina, but in compressed form according to certain principles. The compressed form of the cosmic panorama can be called reality seen in perspective. The principles of compression of the cosmic panorama are the principles on which perspective works. Consequently, in order to understand cosmic phenomena it is imperative to understand the principles on which perspective works. One of the principles of perspective is that surfaces far away from the observer are perceived by the human eye as concave or convex.

VII. The Visual Matrix

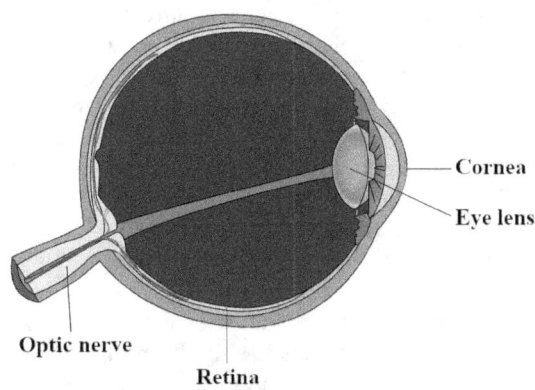

When I refer to the eye, I am actually referring to the emmetropic eye, i.e. the eye that produces normal vision. Light enters through the cornea, through the other layers of the eye and through the lens. The lens is a biconvex lens. Finally, the image outside is projected onto the retina in inverted form. The retina is a layer no more than 0.5 mm thick. This layer consists of photosensitive cells that is located on the inner side of the eyeball wall. In adult humans, the retina is about 72% spherical. This approximate sphere is about 22 mm in diameter.

Because the retina is approximately spherical and the image is projected onto the concave side of the retina, the visual system always produces a concave image. Sometimes, due to the structure within the image and the varying degrees of brightness of the colours within it, concavity is perceived to be convexity. Cells on the retina convert the image into nerve impulses. These impulses are transmitted via the optic nerve to the brain. The brain interprets the impulses through a complex mechanism made up of millions of neurons. All the lights in the sky such as the Sun, the so-called Planets, the Moon, the Stars, the Sun's Halo, the Moon's Halo, the

Rainbow and so on appear to be circles. The eye transforms the light coming from these phenomena into circles. The fact that the eye always produces a circular and concave image for surfaces far from the observer can be seen by looking at the sky over a longer period of time. If you observe the sky over a long period of time, and especially when there are many clouds in the sky with interstitial spaces between them, you will notice that the sky appears to be a dome. The sky appears to be a dome because the earth's atmosphere is perceived by the eye as a concavity.

Transformation of shapes into round shapes by the retina: The Earth's atmosphere consists of the troposphere, stratosphere, mesosphere, thermosphere and exosphere. The exosphere is the atmospheric layer at the highest altitude and is where the atmosphere thins out and merges with a space called interplanetary space by so-called academics. The term atmosphere is made up of two words, the Greek atmos and the word sphere. Atmos means: vapour. Consequently, the term atmosphere means: the spherical gaseous envelope surrounding the Earth. All other atmospheric substrates are also regarded as spheres. There are many other words ending with the word sphere such as: ozonosphere, biosphere, magnetosphere, hydrosphere, ecosphere, lithosphere, ionosphere, noosphere, photosphere, thermosphere, chromosphere, etc. It should be noted that all these phenomena described by the above terms are phenomena extending over a very large spatial area. They usually extend over an area of terrestrial dimensions or even larger. From this we can see that the phenomena described by the above terms appear to be spherical. This is why these terms end with the word sphere. People have the impression that these phenomena are spherical due to the fact that the image is projected onto the concave side of the retina and that the retina is about 72% spherical in shape which makes the distant features appear to be concave or convex. After the eyes transform them into convex entities, the mind interprets them to be spheres or domes.

VII. The Visual Matrix

If you look at a plane flying in a straight line and always at the same altitude and at a very long distance from you, you will notice that it appears at the horizon line and will make an upward curve and then make a downward curve and disappear behind the horizon line even though it is actually flying in a straight line. The plane's straight path is transformed by the retina into a curve. **If** you are in a long tunnel with a square-shaped entrance, you may notice that as you move away from it, the shape of the entrance tends to become rounder and rounder as the distance between you and the entrance increases. **If** you stand back from a cogwheel and look at it, you'll notice that the wheel's teeth disappear little by little until you finally get the impression that the wheel is round like a disc. **If** you look at an ellipsoidal object from a great distance it will appear to be round. You can simulate this experience on your computer. Use a graphics program to draw a white ellipse on a black background. It is good if the ellipse is white and the background is black to see the ellipse better. Use the magnifying glass to make the ellipse smaller. You will notice that when the ellipse shrinks, it tends to morph into a circle.

Because the degree of ellipse decrease with a graphics program is much smaller than the degree of decrease of objects at Earth or cosmic distances, you can see the transformation of the ellipse into a circle better if the ellipse is shaped more like a circle. Luminous shapes change into round shapes at smaller distances than the distances at which material objects change into round shapes.

Mind: The mind works on division. In other words, the mind understands a phenomenon better and faster if the phenomenon to be examined is divided into its subcomponents. In order to understand the perspective mentally better and faster I will divide it into its subcomponents. The subcomponents of perspective are: planetary perspective, aerial perspective and cosmic perspective.

The planetary perspective: The planetary perspective is the one that an observer on land or water has when looking towards the horizon. It can be at sea level, in which case sea level is considered to be level zero, or at higher altitudes such as from a mountain. In the planetary perspective, the land or water between the observer and the horizon appears to rise from the observer towards the horizon, while the sky appears to descend towards the horizon. The horizon can always be seen from the ground or on water provided it is not obstructed by any obstacle and the atmosphere is not foggy.

Because the earth appears to rise towards the horizon and the horizon is all around you, you can see that the horizon is actually a circle whose surface is perfectly horizontal. All the lines below the horizon line that are parallel to it describe circles that get smaller and smaller the lower they are below the horizon line. This is why the entire surface around the observer that lies between the observer and the horizon appears to be concave even without looking at it from the basket of a hot air balloon. From the basket of a hot air balloon, the appearance of concavity of this surface is more perceptible because

VII. The Visual Matrix

you can look at it from a high altitude and because there are no obstacles in front of you like on Earth to obstruct your horizon, which makes you perceive the horizon as a circular line around you. The higher the observer's eyes are, the more land or water they can see.

The distance to the horizon line can be calculated with the following formula:

$$d \approx 3{,}57 \sqrt{h}$$

d = Distance in km to the horizon
h = the height from the ground to the observer's eye, expressed in metres

In the following image is a photo taken in a long corridor. The white and oblique lines represent the main sight lines. All sight lines intersect at a point. The point where the sightlines intersect is called the vanishing point. The floor of the corridor appears to rise from the observer towards the vanishing point, while the ceiling appears to descend from the observer towards the vanishing point. The white horizontal line is the line at eye level. It corresponds to the horizon line in nature.

Not only does the floor seem to rise, but also the sight lines on the side walls below eye level. Everything below eye level seems to incline in an upward direction. Not only does the ceiling seem to descend, but also the sight lines on the side walls above the eyes. Everything above eye level seems to descend. In summary, everything below the line at eye level appears to incline in an ascending way, and everything above the line at eye level appears to incline in a descending way. Now to an example from nature. In the following picture we can see the horizon line, a white line and a gold line. The white line is parallel to the horizon line and lies below it. The gold line is parallel to the horizon and lies below the white line. Notice that the surface of the water appears to be inclined upwards! This means that the surface of the land will also appear to incline upwards. Notice that the Earth's atmosphere appears to incline downwards!

VII. The Visual Matrix

If you could look around, as you might from the basket of a hot air balloon, you might notice that the horizon appears to be circular and all around you. The plane formed by the horizon is a horizontal plane. The white line in the picture below the horizon is parallel to the horizon. Like the horizon line, this line appears to be circular, but the circle formed by it is slightly smaller than the circle formed by the horizon line. The gold line below the first white line is also parallel to the horizon. This line also appears to be a circle.

The circle formed by this line is slightly smaller than the circle formed by the white line. If you were to draw many lines parallel to the horizon line and placed lower and lower from it, then you would notice that all these lines describe a concavity. All around you, within a few metres or so you will be able to perceive the ground as flat if you are on level ground. From this example, you can tell that the land area demarcated by the horizon line appears to be concave and therefore looks like an upside-down bowl or an upside-down dome.

The same is true of the space above the horizon line. The higher the lines above the horizon line are above the horizon line, the smaller the circles they form. This means that the Earth's atmosphere looks like a dome. This is why some people claim that the planet is covered by a dome. The apparent concavity of the ground and the dome shape of the Earth's atmosphere, however, are due to the eye transforming visual lines below the horizon and above the horizon into circles. Let's now look at what happens to distant objects and those that disappear out of sight.

Imagine that the sea craft in the picture is moving away from you, and you are standing on the shore of that water! The further the boat moves away from you, the smaller and smaller it will seem until it disappears from your field of view. But with the help of a telescope, you will be able to bring the craft back into your field of view. In

VII. The Visual Matrix

other words, with the telescope you will be able to see that craft. This is very easy to understand.

There is, however, another aspect that many people do not understand, and because of this they make wrong inferences about the interpretation of perspective. You can bring this craft back into the field of view with a telescope because it didn't cross the horizon. What crosses the horizon line will apparently start to sink. The submerged portions can no longer be brought back into the field of view with a telescope from sea level. To bring the submerged portions back into the field of view you must use the telescope, but in addition you must increase the viewing altitude until the observed object is between you and the horizon. By increasing the viewing altitude, the length of the field of view between you and the horizon increases so that the observed object can be brought between you and the horizon. When the observed object is between you and the horizon, you will be able, using the telescope, to bring the observed object back into your field of view so that you can see even the parts that appeared to be sunken.

In photo 1 in the following image you can see the buildings of Chicago from the opposite shore of Lake Michigan and from about 60 miles away.

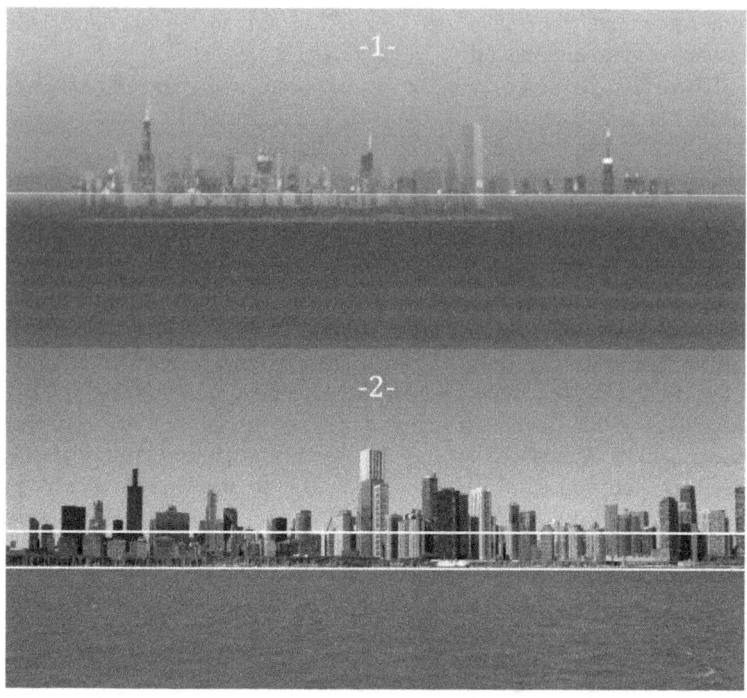

These buildings have shrunk a lot because of the distance between the observer and the buildings. Because these buildings are large, they have not become so small that they disappear from the field of view. What's more, they crossed the horizon and began to sink. The sunken portion is below the horizon. The portion below the horizon can no longer be brought into view with the telescope from sea level. In order to bring this portion back into the field of view with the telescope, you must increase the viewing altitude until the sunken portion is between you and the horizon. In image 2 you see the same buildings from such a short distance away that they can be seen

VII. The Visual Matrix

shrunk, but whole, so not sunken. From the sinking phenomenon of objects passing beyond the horizon, we can deduce that the Earth appears to incline descendingly on the portion beyond the horizon. This leads to the inference that the Earth's atmosphere appears to incline upward on the portion beyond the horizon.

Aerial perspective: The aerial perspective is at an altitude of about 3-4 km. In the aerial perspective, the planet and sky appear to be concave, forming a lens-like shape together.

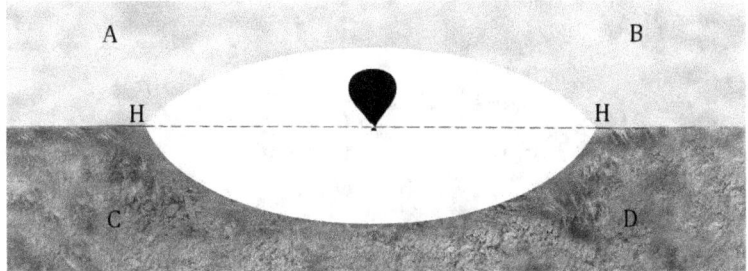

The image above shows this lens shape as seen by an astronaut in a light gas balloon. This astronaut has the perception that he is inside that lens and cannot see outside that lens. AB represents the sky. CD represents the surface of the Earth. HH is the horizon. The horizon line is the line formed by the apparent joining of the Earth's atmosphere to the Earth. It is always at eye level. The horizon line rises or falls with the observer's eyes.

Below the balloon, the surface CD appears to be concave and appears to rise towards the horizon line HH where it appears to merge with the sky.

English meteorologist, astronaut and astronomer James Glaisher wrote in the British newspaper **The Leisure Hour** that the plane of the earth appeared as a concave surface resembling an upside-down watch glass, and the horizon line was the edge of this concavity. The blue atmosphere above it, he goes on to say, appears to be a spherical cap resembling the earth's concavity, but upside down, sitting on the earth's concave surface. Accordingly, the horizon line is circular, lies all around the observer, and results from the apparent conjunction of the Earth's atmosphere with the Earth.

Cosmic perspective: The cosmic perspective is where you can see cosmic phenomena such as the Sun, Moon and Stars.

In this context it is very important to note that the image is projected onto the concave side of the retina and that the retina is about 72% spherical in shape which means that the image is always concave. Consequently, the human eye always transforms bright and illuminated entities in the distance into circles, and the mind interprets them as spheres or domes. Light is of two types: emitted light and reflected light. Emitted light is light that comes from a light source such as the Sun. Reflected light is light that comes from bodies illuminated by a light source such as the surface of the earth and the atmosphere. The human mind sometimes perceives concavity as convexity. Here's an example!

VII. The Visual Matrix

In the image below is a piece of art shown in three positions. This piece of art contains a convex face and a concave face. In positions A and C, so those in lateral-frontal view, it can be seen that on the left there is a convex face and on the right there is a concave face. In position B, the artwork is viewed from the front and it appears that both faces are convex. The convex appearance of the face on the right is also called the Hollow Face illusion. If you move to the left or right across the face of the Hollow Face, you get the impression that the face is moving to the left or right, following you with its eyes. The convexity of celestial entities arises due to an illusion similar to the Hollow Face illusion.

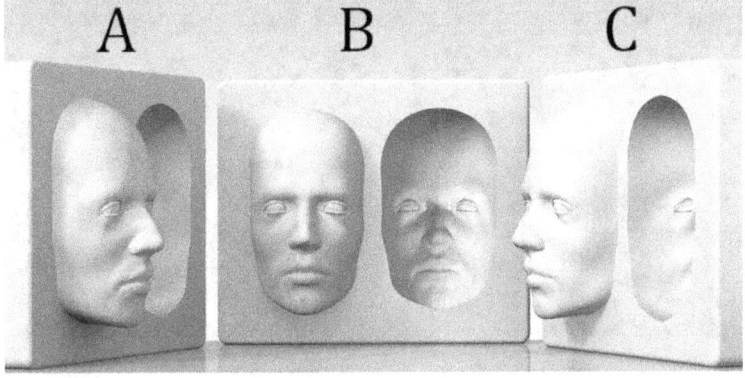

Images in the shape of a circle containing emitted light are perceived as convex. Circle-shaped images containing reflected light are generally perceived to be concave. This is why the Earth and Earth's atmosphere, seen from above, for example from a hot air balloon, appear to be concave, and the Sun, Moon and Stars appear to be convex. Because the Sun, Moon and Stars appear to be convex, the human mind interprets them to be spherical or approximately spherical in shape.

All the lights in the sky, such as the Sun, the so-called Planets, the Moon, the Stars, the Sun's Halo, the Moon's Halo, the Rainbow appear to be circles. The eye turns the light from these phenomena into circles. The fact that the eye always produces a circular and concave image of distant entities can be seen by looking at the sky over a longer period of time. Especially when the sky is full of clouds and there are interstitial spaces between the clouds you will notice how the sky appears to be a dome. The sky appears to be a dome because the earth's atmosphere is perceived by the eye as a concavity. In the next image you can see a solar halo.

In the next image you can see a lunar halo.

VII. The Visual Matrix

The cosmic perspective: The following picture shows the cosmic perspective. The observer is on land at sea level. The O points are two points on the circle formed by the horizon line. The horizon is therefore a circle around the observer. The sky is represented as a dome because that is how the observer's eyes see it. The Earth is represented as a plane, so as it really is. The line of sight is the line from the observer's eye level perpendicular to the horizon. The line of sight appears to be tilted towards the Earth because the Earth appears to be rising. At the same time, it seems to be tilted towards the sky due to the fact that the Earth's atmosphere seems to be descending. Since the Earth is represented as flat in this graph, the line of sight is tilted towards the Earth until point O, after which it is tilted towards the sky. Notice that part of the sky cannot be seen.

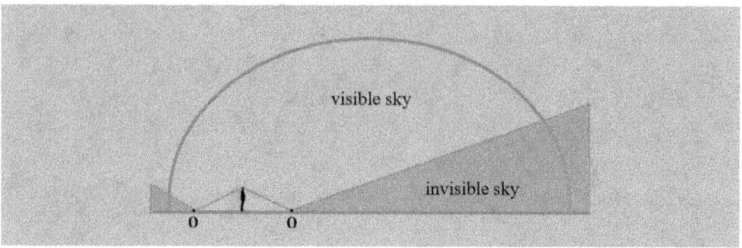

How much of the sky is visible depends on where the observer is standing and which direction they are looking. From this graph it can be seen that the observer cannot see the whole planet, but only the area of land that is surrounded by the horizon. At the same time, he sees an area of sky that is much larger than the area of land formed by the horizon.

Structure of the Perspective: The structure of the perspective is infinitely complex. In this graph we have represented the structure of the perspective in a much simplified way so that the likelihood of the average person understanding it is as high as possible. For example, I have drawn straight lines instead of curves. Note, however, that some lines on the graph are seen by the eye as curves. The Earth's tilt, that of the Earth's atmosphere and that of the cosmic atmosphere are curves. Consequently, the upward tilt of the Earth from the eye to the TH point appears to be concave. This concavity is oriented upwards, and the downward tilt of the Earth and Cosmic Atmosphere also appears to be concave. This concavity faces downwards. The TH, CH and KH points are actually perpendicular lines on the plane of the graph. The line from eye level through the TH-CH-KH points is actually a plane perpendicular to the plane of the graph. The line of sight is the line from the midpoint of the distance between the two eyes perpendicular to the horizon line. Notice the seemingly paradoxical fact that the line of sight appears to be tilted towards the Earth because the Earth appears to be rising. At the same time, it

VII. The Visual Matrix

appears to be tilted towards the sky because the Earth's atmosphere appears to be descending.

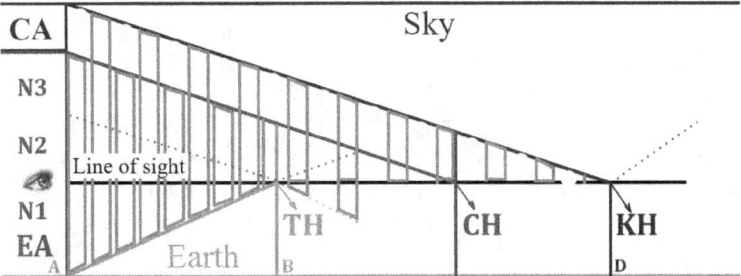

CA = Cosmic atmosphere
EA = Earth atmosphere

TH = The terrestrial horizon
CH = The celestial horizon
KH = The cosmic horizon

The Earth is perfectly horizontal, but it appears to tilt upwards from the observer towards the horizon. Earth's atmosphere is perfectly horizontal, but it appears to tilt downward from the observer towards the horizon. Point TH is actually a line perpendicular to this plane and represents the terrestrial horizon. The sky is made up of the Earth's Atmosphere (EA) and the Cosmic Atmosphere (CA). The EA is made up of three layers at different levels: the first layer (N1), the second layer (N2) and the third layer (N3). N1 consists of the terrestrial atmosphere below the observer's eyes. N2 consists of the terrestrial atmosphere above the observer's eyes. N3 consists of the terrestrial atmosphere above N2. N1 tilts upward to the TH level. N2 tilts downward to the TH level. N3 tilts downward on the surface of the sky between TH and KH. On the TH-KH portion the earth appears to tilt downward. This is why large objects that don't disappear completely out of view until the horizon line seem to sink in. Over the distance between TH and CH, N3 appears to tilt upward which makes it overlap by self-projection on the sky. We can say that over the distance between TH and CH, N3 self-projects into the sky. Over the distance between CH and KH, the Cosmic Atmosphere appears to tilt downwards. From the point KH onwards, the sky tilts upwards which makes it self-projecting on itself. From the KH point onwards, the Sun is no longer in the sky which means that a black disc will project into the sky instead of the Sun. The ancients called it the Black Sun. Briefly put, the sky consists of three superimposed layers: the real atmosphere and two more projections of it.

The first projection of the real atmosphere, i.e. between TH and CH, I will call the Primary Projection (PP). The second projection of the real atmosphere, i.e. after KH, I will call the Secondary Projection (SP). Both PP and SP are holographic projections of the real atmosphere.

On the sky you can project holographic images. These holographic images can be seen in many situations when holograms appear in the sky in the form of buildings, pyramids, crosses and other shapes. These holograms are actually light from various large objects that are projected onto the Earth's atmosphere. Many people believe that the holographic images projected on the sky are divine signs confirming their religion, others believe that they are just fake photographs or that they are projected on the sky by someone. However, many of them are real and are projections of the Earth's atmosphere.

Those who are not familiar with the workings of perspective are of the opinion that if the Earth were flat then at sunset, the Sun would have to become so small on the horizon that it would have to

VII. The Visual Matrix

disappear from the field of view, but could be brought back into the field of view with the telescope. This view is totally wrong. What happens to buildings that sink beyond the horizon also happens to the Sun. The Sun, being a large object in the sky, shrinks to the horizon, but does not become so small that it disappears from the field of view. It passes beyond the horizon in the shape of a circle and slowly, slowly begins to sink as the distance between the observer and the Sun increases.

After the Sun disappears completely below the horizon, this phenomenon is called sunset, it can no longer be brought back into the field of view with a telescope because, as I said before, sunken objects can no longer be brought back into the field of view with a telescope from sea level. Submerged objects behind the horizon, which are on the interval between TH and CH, can be brought back into the field of view with the telescope and by increasing the altitude until they return between the observer and TH. The Sun begins to sink in the interval between TH and CH. While the Sun is in the interval between TH and CH, it can be brought back into the field of view with a special telescope, made to be eye-friendly, but only from such a high altitude that it returns between the observer and TH. At sunset, the Sun finally sinks completely behind KH. When the Sun sinks completely behind KH, it cannot be brought back into the field of view in any way.

VIII. Tellus Plana

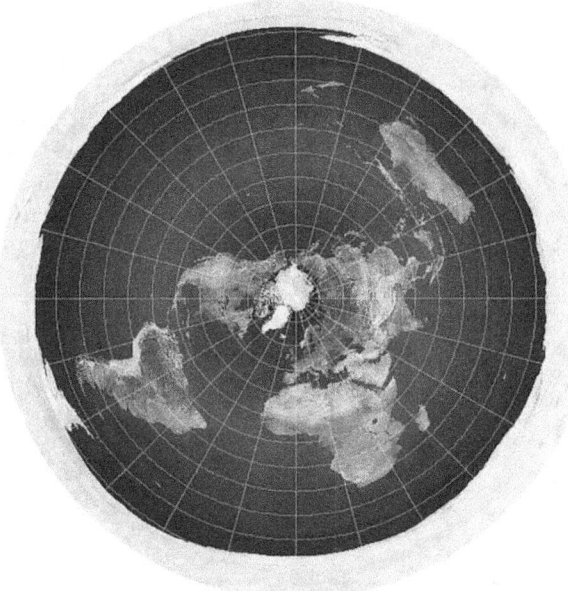

THE EARTH IS FLAT.

Flat Earth refers to the Earth's surface when we ignore landforms such as mountains, volcanoes, hills, valleys, ripples on the water surface and crust on the Earth's surface. The Earth's surface crust is generally sloping like this: at the equator it is highest and slopes down towards the North Pole and Antarctica. This is why most of the rivers on the southern hemiplane flow south and most of the rivers on the orthern hemiplane flow north. There are some streams that lie on the southern hemiplane and drain to the north, and others that lie on the northern hemiplane and drain to the south. These streams lie on a specific sloping crust leading to the mouth of that stream. If the Earth were perfectly flat, the water would collect around the outflow, but it would not flow as a stream, such as a river. Consequently, there is a crust on Earth, and on this crust is the planetary relief. The official paradigm, i.e. the self-styled scientific community, supports the convex-heliocentric theory that the planet is a convex-shaped object with a radius of about 6,373 km. To understand that the planet is flat and not convex, it is good to understand the difference between a vertical on a flat planet and a vertical on a convex planet. To define the two vertices and express them geometrically, we need to idealize the flat planet and the convex planet and transform these two entities into a plane and a sphere.

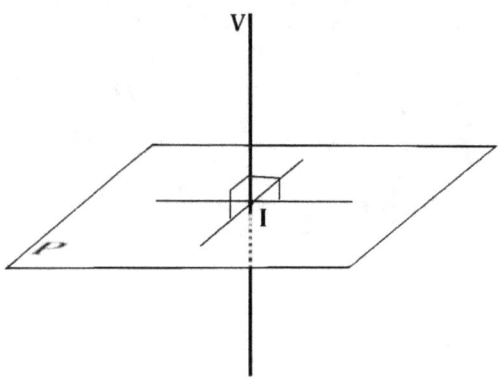

Let plane P and vertical V be given. The point of intersection between the vertical V and the plane P is point I. Vertical V is perpendicular to all lines on plane P that pass through point I. Generalised, a line V is vertical on a plane if it is perpendicular to at least two lines on that plane that pass through the point of intersection of line V with that plane.

VIII. Tellus Plana

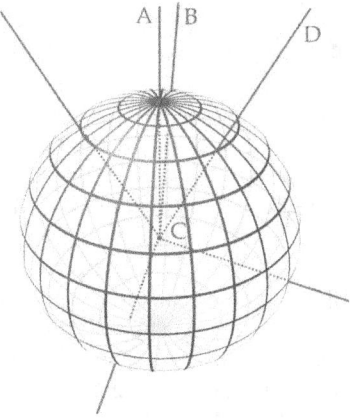

On a sphere all lines passing through the centre of the sphere are the vertices of that sphere. Notice that vertical B is tilted backwards from vertical A, and vertical D is tilted forwards and sideways from vertical A! If the planet were convex, objects on the ground, in the water and in the air at great distances, such as buildings, ship masts, hot air balloons in the air, etc., would appear tilted. This has never been noticed, which means the planet is flat.

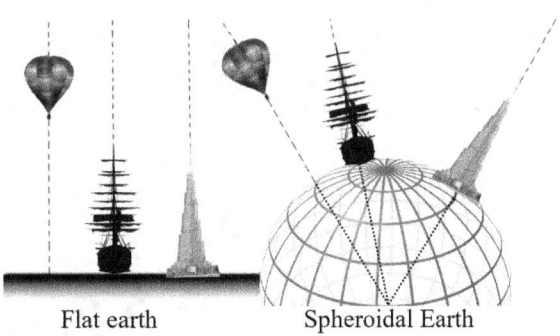

Flat earth   Spheroidal Earth

You can find out the angle of inclination of an object relative to the vertical of the place where you are by using the following formula:
α = d/r.
r = radius of the planet = 6,373 km.
d = distance measured on the ground to the observed object.
α = the angle of inclination from the vertical of your location.

Even at small angles of inclination, the tilt of objects on the ground, on water or in the air should be visible because the radius of the planet is very large and the tilt of an angle becomes more pronounced the further you are from the apex of that angle. On a convex planet, the apex of the tilt angle is in the centre.

Calculation of the height hidden by the curvature of the convex planet: The surface of standing water, for example the surface of lakes, seas and oceans on the convex planet must be convex and therefore every part of this planet must be an arc. From the apex of each arc of a circle there must be a curvature with a declination (= fall = declivity) of 8 cm for the first km, 32 cm for the second km, 72 cm for the third km and so on, the rule being that the declination is the result of multiplying 8 by the square of the distance expressed in km. This rule must be changed after 1 600 km. The following picture explains this. In the following image, the observer's eyes are at point T.

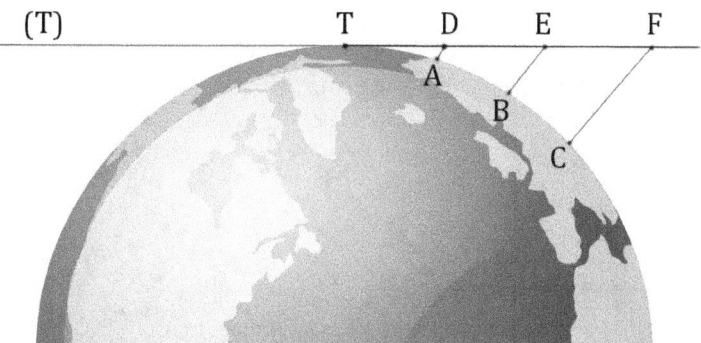

VIII. Tellus Plana

The declination at point A is the distance between point A and point D. In this case, the observer can only see from point D upwards. The declination at point B is the distance between point B and point E. In this case, the observer can only see from point E upwards. Let's find out the formula for calculating the declination, which we can also call the height hidden by the curvature of the convex planet, if the observer's eyes are at a certain height! Assumed starting data:

.the Earth is spherical.
.The radius of the planet is 6,373 km long.

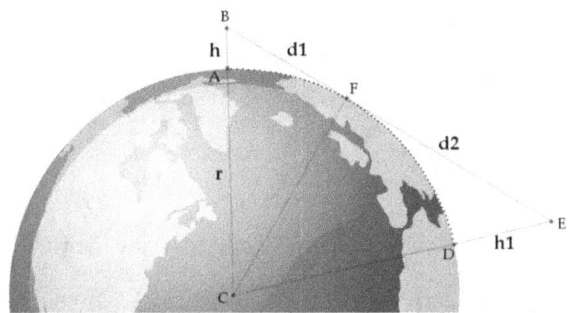

Suppose the observer is at point A on the surface of the convex planet. Suppose a tall object is at point D. If the object is less than height h1, we can no longer see the object because of the curvature of the convex planet. If the object is taller than height h1, the part of the object from the ground to point E we will not see.

We want to find the hidden height $DE = h1$ to do some experiments to see if the planet is curved or not.

C = centre of the Earth.
$CA = CF = CD = r$ = Earth's radius.

AB = h = height from ground to observer's eye.
BE = d = distance between observer and object to be observed.
BF = d1.
FE = d2.
ED = h1 = height hidden by the Earth's curvature.

    We know the elements h and d. We want to find the height hidden by the curvature of the convex planet, so we want to find the height h1 mathematically. We can find this height using the right triangles CBF and CEF. In the first step we find the distance d1 from the right triangle CBF.

$(r + h)^2 = r^2 + d1^2 =>$
$r^2 + h^2 + 2rh = r^2 + d1^2 =>$
$d1^2 = r^2 - r^2 + h^2 + 2rh =>$
$d1^2 = h^2 + 2rh =>$

$$d1 = \sqrt{h^2 + 2rh}$$

In the second step we find the height h1 using the right triangle CEF.

$(r + h1)^2 = r^2 + d2^2 =>$
$r^2 + h1^2 + 2rh1 = r^2 + d2^2 =>$
$h1^2 + 2rh1 - d2^2 = 0$

This is an algebraic equation of the second degree. We find the height h1 by extracting the square root from this equation.

$$h1 = \sqrt{r^2 + (d-d1)^2} - r$$

With this formula we can calculate the height hidden by the curvature of the convex planet to do experiments by which we can find out whether the planet is curved or flat.

VIII. Tellus Plana

Samuel Birley Rowbotham (1816-1884) was an English inventor and writer who wrote the book Zetetic Astronomy: Earth Not a Globe under the pseudonym Parallax. To write this book, he studied the planet for decades. The term zetetic comes from the Greek verb zeteo which means: to examine, to investigate. In ancient Greek, the word zetetic became a technical term for the endeavour to know through research and investigation. In his book, Samuel Birley Rowbotham presents several experiments he did to prove that the planet is flat and not convex. In the following I present three of those experiments.

Experiment 1: In the county of Cambridgeshire in England there is an artificial river, which is actually a canal, called Old Bedford. That river is more than 20 miles (≈32 km) long and flows in a straight line through a part of the plain called Bedford Level.

Its water is mostly still and quiet. Along its entire length the river is uninterrupted by locks and weirs. These are the reasons why that river is in all respects very suitable for the purpose of ascertaining whether the upper surface of the river is vaulted, and of ascertaining the extent of its convexity, if any. A boat was sent to sail from a place called Welney Bridge to another called Welche's Dam. That boat was equipped with a flag. Welney Bridge is 6 miles (≈ 9.65 km) from Welche's Dam. The observatory was in Welney Bridge and stood in the water with a good telescope. The distance between the observer's eyes and the water level was no more than 8 inches (≈ 20.32 cm). If the planet had been convex, the river water would have been 6 feet (≈ 1.82 m) higher in the center of the 6-mile (≈ 9.65 km) long arc than the two locations, Welney Bridge and Welche's Dam. The observer's eyes were 8 inches (≈ 20.32 cm = 0.2032 m) above water level. Let's calculate the approximate distance to the horizon line with the formula:

$$d \approx 3{,}57\sqrt{h}$$

d = distance in km to the horizon.
h = height from ground to eye measured in metres = 8 inches = 0.2032 m.
d ≈ 3.57 x 0.4507 ≈ 1.6093 km ≈ 1 mile.

The distance to the horizon line is the distance between the observer and the point of tangency T between the water surface and the line of sight. After point T, the water surface should have dropped below this point by 16 feet (≈ 4.88 m) and 8 inches (≈ 0.20 m) at the end of the remaining distance of 5 miles (≈ 8.05 km). This means that she would have had to descend by 5.08 m (4.88 m + 0.20 m = 5.08 m). You can find out the water's declination (hidden height) with the following formula: $8 \times 5^2 = 200$ inches (≈ 5.08 m).

The curve AB represents the arc formed by the convexity of the water on the convex planet. The line L is the line of sight and is tangent to the arc AB at the point of contact T. The distance between the point of contact T and the observer's eyes at point A is one mile (≈ 1.60 km). The distance between contact point T and the boat at point B is 5 miles (≈ 8.04 km). To find the height hidden by the curvature of the convex planet, i.e. to find the level of the boat above the water surface, multiply 8 inches (≈ 20.32 cm) by the square of 5: 8 x 25 inches = 200 inches (≈ 508 cm = 5.08 m). The hidden height is 5.08 m. Therefore, the boat should have been 5.08 m below the water

VIII. Tellus Plana

level at point T. The flag was 3 feet (≈ 0.91 m) high and therefore should have been about 13 feet (≈ 4.17 m) below the line of sight L.

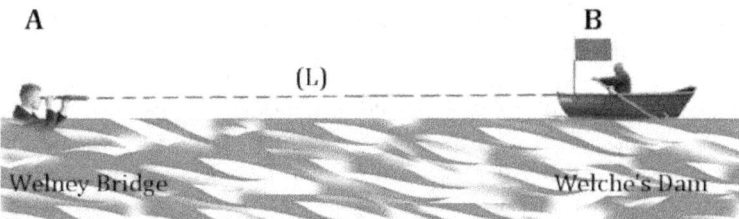

The flag and the boat were clearly visible at all times along the entire route the boat travelled. The conclusion was that the water level did not drop any from the line of sight. It follows from this experiment that the surface of a still water is not convex and therefore the planet cannot be convex. It follows that the planet is flat. The surface of the water looks like the picture above. The water level is horizontal and parallel to the line of sight.

Experiment 2: Along the water's edge of the Old Bedford Canal, 7 flags were placed one mile (≈ 1.60 km) apart. The flags were arranged so that the tip of each flag was 5 feet (≈ 1.52 m) from the water surface. Near the last flag, a rod carrying a flag with an area of 3 square feet (≈ 0.2787 m²) was attached.

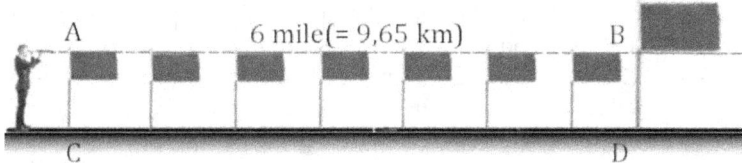

The top tip of this rod was 8 feet (≈ 2.44 m) above the water level. The bottom edge of this flag lies on the line of the top edge of the other flags as in the picture. Looking through a good telescope over and along the row of flags from point A to point B, the line of sight lies along the bottom edge of the large flag at point B. The elevation of point B above water level at point D was 5 feet (≈ 1.52 m), and the elevation of the telescope at point A above water level at point C was also 5 feet (≈ 1.52 m). All intermediate flags had the same altitude. Since the water surface represented in the drawing by the line CD was equidistant from the line of sight AB, and AB was a straight line it follows that CD was parallel to AB and therefore CD is a straight line. In other words, the surface of the water, hence the line CD, was absolutely horizontal for a distance of 6 miles (≈ 9.65 km). If the planet were convex, the series of flags would be on a curve like in the following picture.

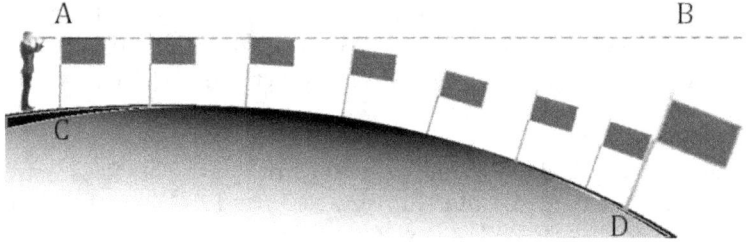

In the case of a planet with a convex surface, because the surface of the water represented by the CD curve would be domed, each intermediate flag would be below the line of sight represented by the AB line. The first and second flags would have determined the line of sight AB, and the third flag would have been 8 inches (≈ 20.32 cm) lower than the second flag. The fourth flag should have been 32 inches (≈ 81.28 cm) lower than the second flag. The fifth flag should have been 6 feet (≈ 1.82 m) lower than the second flag. The sixth flag should have been 10 feet (≈ 3.04 m) and 8 inches (≈ 20.32 cm) lower than the second flag. The seventh flag should have been 16 feet (≈ 4.87 m) and 8 inches (≈ 20.32 cm) lower than the second. The top of

VIII. Tellus Plana

the last and largest flag, being 3 feet (≈ 0.91 m) higher than the small flag, should have been 13 feet (≈ 3.96 m) and 8 inches (≈ 20.32 cm) below the line of sight at point B. The curvature of a convex planet would have required that flags below the line of sight could not be seen. The flags, however, could be seen. It follows that the convex-heliocentric theory is a mere theory that cannot be supported by scientific and experimental evidence. The convex-heliocentric theory is an invention of abstract and ignorant minds. These two experiments were carried out by Samuel Birley Rowbotham in the summer of 1838.

Experiment 3: A good theodolite was located on the north shore of the Old Bedford Canal. It was located midway between Welney Bridge and Old Bedford Bridge. This distance is 6 miles (≈ 9.65 km) long. The line of sight of the horizontally mounted theodolite lies at the elevation of line BB.

Each point B lies at a distance of 3 miles (≈ 4.82 km) from point T. If the planet were convex, each point B would be 6 feet (≈ 1.82 m) below the line of sight as shown in the following image.

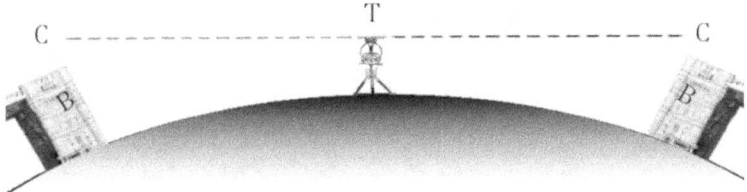

However, the B points lie on the line of sight which means that the planet has no degree of curvature. In other words, the planet is flat. If you stand on the deck of a sea craft, on the top of a pole, on the top of a mountain or in the basket of a hot air balloon and look off into the distance, the surface of the land and water appear as a vast sloping plane that rises all around you towards the horizon, getting higher and higher as the distance increases until it reaches the horizon. If you rise to a greater height, or if you descend to a lesser height, the horizon line also rises or falls, thus remaining at the observer's eye level at all times. You feel as if you are in the centre of a huge amphitheatre. You feel as if you are in a concavity whose edges expand if you rise to a greater height, or contract if you descend from that height.

In the book A System of Aeronautics, John Wise (24.02.1808 - 28.09.1879), an American aeronaut, says: „The apparent concavity of the Earth as seen from a balloon. The visible plane below describes a perfect circle, I might rather call it a concave-sphere, for we have reached a height from which the Earth acquires an appearance of a cavity or concavity - an optical illusion which increases as you move away. At the highest altitude I reached, which was about a mile and a half (1.5 miles ≈ 2.4 km), the earth around me seemed to have an appearance or shape like that made by joining the edges of two watch glasses, and the balloon seemed to be in the center of the cavity throughout the flight at that altitude." The image below shows this appearance.

VIII. Tellus Plana

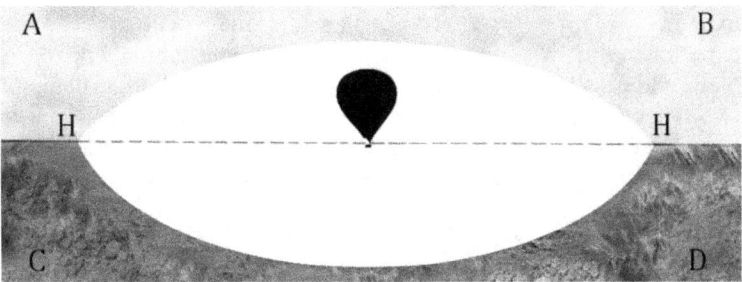

AB represents the sky. CD represents the surface of the earth. HH represents the horizon. Below the balloon, the surface CD appears to be concave and appears to rise towards the horizon line HH where it appears to merge with the sky. The horizon line is always at the observer's eye level. It rises or falls with the observer's eyes.

In the following picture you can see the case where the planet would be convex.

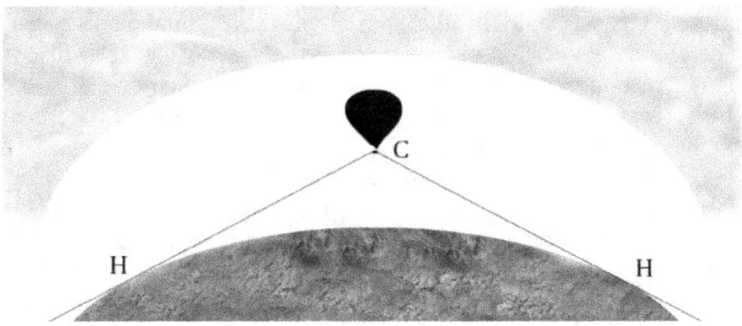

If the planet were convex, the horizon line HH would be convex and would descend continuously as the balloon rose to a higher altitude. You can understand in a very simple way that the planet is flat and not convex by looking at the image below and performing an experiment. Before I go on to explain the image and describe the experiment, I want to explain some aspects of light and the horizon.

In the next few lines I will tell you a little about light. There are two types of light: emitted light and reflected light. Emitted light is light coming directly from a light source. Sunlight and candlelight are emitted light. Through emitted light you cannot look, no matter how dim that light is. You can't even look through a candle flame.

Reflected light is light coming from a surface illuminated by a light source. Imagine you're standing in the dark in front of a wall and you have a flashlight with you! If you switch on the flashlight and shine it on the wall in front of you, you will be able to see the piece of wall illuminated by the flashlight. In this way you will see the structure of that piece of wall. You can see the structure of that piece of wall because reflected light is coming towards your eyes from that piece of wall. The light coming from the wall is reflected light. Light reflected from a surface reveals details of that surface to you. In the next few lines I explain something about the horizon. Let's imagine there was a convex planet! The horizon line of a convex planet would be the convex profile of that planet itself. Therefore, on a convex planet, at sunrise, if most of the solar disc is above the horizon line and the smallest part of the solar disc is below the horizon line, the horizon line is very bright because it is formed by the light emitted by the Sun.

On the flat planet, the horizon is formed because the human visual sense sees reality in perspective. In perspective, the terrestrial plane appears to rise and the celestial plane appears to fall, so that the intersecting line of the two planes appears at eye level. This intersecting line is the horizon line itself. In this case, the light coming from the horizon line is light coming from the two planes and is therefore reflected light. Reflected light is much weaker than emitted light.

I will now go on to analyse the illustration composed of image 1 and image 2. Image 1 is produced on a computer. This image shows a

VIII. Tellus Plana

sunrise on a convex planet. As you can see, in this case the horizon line is curved and is very bright. This horizon line is curved because it is the actual profile of the planet, which in this case is convex. This horizon line is very bright because it is an emitted light coming from the Sun. Image 2 is an image produced by photographing reality. This image also contains a sunrise. As you can see, in this case the horizon line is a straight line and is not bright. The horizon line is a straight line because it comes from the intersection of two planes, namely the intersection of the terrestrial and celestial planes.

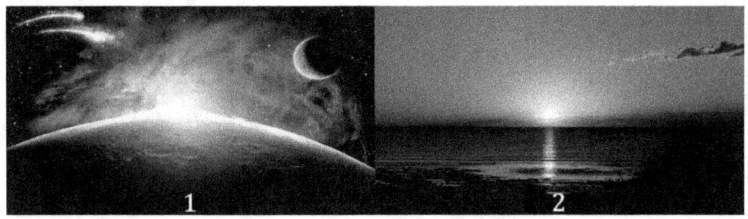

The horizon line is not bright because it is reflected light. The sun illuminates the terrestrial plane and the celestial plane, and the horizon is the line reflected from the intersection of the two planes. Consequently, image 1 does not correspond to reality and therefore the planet cannot be convex. I will now go on to explain the experiment by which you will be able to understand for yourself and through experimentation that the earth is flat and not convex. Before explaining the experiment it is good to look at the above pictures once more. In picture 2 you can see how the Sun produces a streak of light on the water. This streak of light is on the water and the upper surface of the water is always flat.

This streak of light looks like a much elongated trapezoid with a large base near the observer. If the observer is standing on the ground at sea level, then the distance from the observer to the horizon is about 5 km and so the streak of light produced by the Sun on the surface of the water is about 5 km.

If the observer is standing on a hill or tower 100 metres above sea level, then the distance from the observer to the horizon is about 36 km and so the streak of light produced by the Sun on the surface of the water is about 36 km. Such a streak of light can only originate on a flat surface, but cannot originate on a convex surface. To see if this statement is correct you can do the following experiment. To do this experiment you need a flashlight and a flat surface. I assume you chose to do the experiment on the flat surface of a table.

Place the flashlight in such a way that you simulate the rising of the Sun, where a large part of the solar disc is above the horizon and a small part of the solar disc is below the horizon. The edge of the table will act as the horizon line. In image 2 notice that the Sun's surface appears to be vertical, while the Earth's surface slopes upwards from the observer towards the horizon. Consequently, there appears to be an angle of slightly more than 90 degrees between the Sun's surface and the Earth's surface. This is why it is good to tilt the flashlight so that its surface and the surface of the table are at an angle slightly greater than 90 degrees. This will give you a streak of light similar to the one in picture 2. Now try to get a streak of light similar to the one in picture 2 by using a spherical cap instead of the table. You won't succeed! This indicates that the planet is flat and cannot be convex. On a convex planet, the horizon line would actually be the profile of convexity itself. On a flat planet, the horizon line is formed because our visual sense presents reality to us in perspective. In perspective, the land, or water, appears to rise and the sky appears to descend, making the line of intersection of land and sky appear at eye level. This intersecting line is the horizon line itself. The horizon line is a perfect straight line that always appears at eye level.

Airliners fly at an altitude of about 10 km. As you lift off from a convex plane, the convexity profile that would represent the horizon line would descend. However, passengers on airliners can see that the

VIII. Tellus Plana

horizon is always at eye level. This phenomenon is only possible on a flat planet.

When looking through the plane's window, passengers notice that the horizon is curved. The window of an airliner is about 25 cm x 25 cm. If you complete the curvature of the horizon so that you get a circle, you will find that the result is a circle with a diameter of about several metres. Since a convex planet cannot be seen from an altitude of only 10 km as having a diameter of only a few metres, you can tell that the curvature of the horizon line is the result of the aircraft window being concave and not the curvature of a convex planet's horizon line. The planet is flat. If the surface of the planet were convex, as it is assumed to be in the convex-heliocentric theory, the horizon line would not always be at the observer's eye level, but would descend more and more as the plane rises higher in the air as in the following picture.

The picture below shows that on a flat planet, the horizon is always at eye level and is always a perfect straight line, whereas on a convex planet, the horizon should be curved and go lower and lower with increasing altitude.

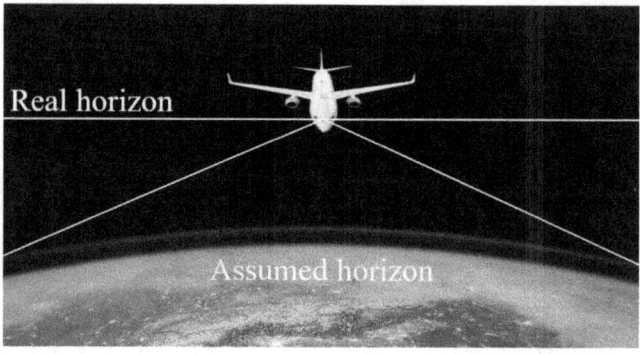

Compass: The compass is a tool used in navigation and orientation. In the case of a flat planet, north is at the centre of the planet, i.e. at the North Pole and below the Polaris star, while the southern extremity consists of all the points on the outer circumference of the planet. The compass points north. The opposite direction to north points south. The plane on which the compass needles move must always be parallel to the plane of the planet, i.e. horizontal. In the case of flat planet, the compass points North and South exactly and simultaneously.

VIII. Tellus Plana

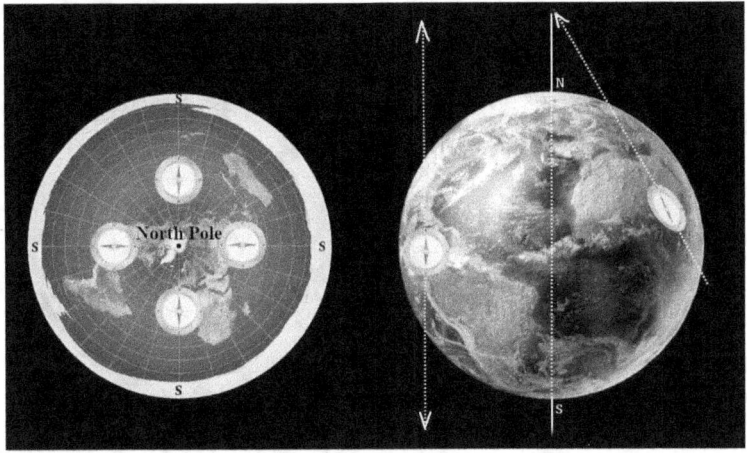

In the convex-heliocentric theory, the planet is convex. Therefore, north would be at the North Pole and south would be at the South Pole. The plane on which the compass needles move must always be perpendicular to the line formed by the centre of the planet and the location of the compass. Consequently, at latitudes near the equator the compass needles will point neither north nor south, but will point upwards towards the sky. On the equator it would be even worse. Those compasses will be parallel to the north-south axis of the planet and will point towards the sky. The maritime compass is a meaningless and impossible instrument to use on a convex shaped planet. The fact that the compass actually works indicates that the planet is flat.

Lighthouses: The Dunkirk lighthouse in southern France is 63 m high and visible from 45 km from a boat, which is 3 m above sea level. The spherical geometry of a spheroid with a radius of 6,373 km, as the Earth is supposed to be, says that the height hidden behind the curvature of the Earth is 118 m, so 55 m higher than the height of this lighthouse. Therefore, according to globular theory, this lighthouse should not be visible from a distance of 45 km. It is visible because the Earth is flat.

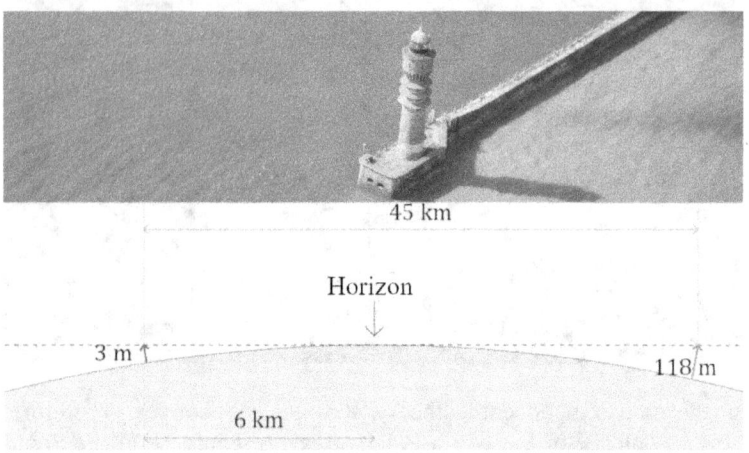

THE EARTH IS IMMOBILE: The convex-heliocentric theory holds that the planet would perform an enormously complex motion consisting of 6 different motions. The first motion would be the rotation of the planet around its axis so that a point on the planet's surface at the equator would have an average speed of 1,675 km/h, which is faster than the speed of sound because the speed of sound is only 1,224 km/h. The second motion would be axial precession, where the planet's axis would spin around another axis, making the planet spin like a peg top which has lost some of its speed. A complete rotation of the axial precession motion would take about 25,772 years. The third motion would be the one of revolution, which would consist of a rotational motion of the planet around the Sun at an average orbital speed of 110,000 km/h. The fourth motion would be the motion of the planet which would be performed as the entire Solar System rotates around the galactic centre. This motion would

VIII. Tellus Plana

be at a speed of 828 000 km/h. The fifth motion would be the motion of the planet that would occur in conjunction with the rotational motion of the Milky Way galaxy around its axis. This motion would be performed at an average orbital speed of 220 km/h. The sixth motion would be the motion of the planet in conjunction with the motion of the Milky Way galaxy in the direction of the constellation Hydra. This motion would be at a speed of 2.15 million km/h.

Star trails: The following picture shows that Polaris, which is the star around which the other stars revolve, and the other stars are above the observer's head in the northern hemiplane, while in the southern hemiplane Polaris and the other stars are above the horizon. From this we can see the following principle: the further away the observer is from the North Pole, the closer Polaris and the other stars are to the horizon.

In one hour, the stars rotate 15 degrees in the sky. Within 24 hours, they make an incomplete rotation. Long-exposure photography of the stars produces star trails that are always concentric and incomplete circles. The trails of stars in the northern hemiplane are always concentric and incomplete circles. At the centre of these circles is the North Star. If the planet were to perform a motion as complex as that described by the convex-heliocentric theory, this would mean that the Pole Star would have to perform all of the mirrored motions of the planet in order to keep itself always in the same position, and the other stars would have to perform, in addition to the mirrored motions of the planet, also their circular motions.

Experiment: A.E. Skellam informs us of the following experiment carried out in England which proves that axial rotation of the planet does not exist. A sturdy cast-iron cannon was placed with the mouth of the pipe upwards. The gun barrel was driven well into the sand up to the muzzle. A piece of fuse was placed at the muzzle. The cannon was loaded with explosive and a spherical cannonball. The barrel of the cannon was thoroughly tested with a plumb line so that it was perfectly vertical. The gun lies in position A. The fuse was lit at point D and the operator retreated to a barrack. The explosion occurred and the cannonball was detonated in the direction AB.

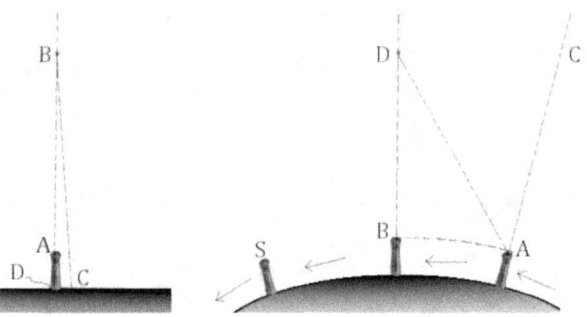

Thirty seconds later, the cannonball fell to the ground in direction BC. The distance between point C and point A was only 8 inches ($\approx$ 20.32 cm). This experiment was repeated several times. Sometimes the cannonball fell right into the cannon's mouth. The largest deviation from point A was less than 2 feet ($\approx$ 0.60 m). The average

VIII. Tellus Plana

time the cannonball was in the atmosphere was 28 seconds: 14 seconds ascending and 14 seconds descending. Consequently, the ground on which the cannon was placed did not move from its position during the 28 seconds the cannonball was in the air.

The convex-heliocentric theory assumes that the planet would be spinning from west to east so that, at the latitude where England is, it would have a speed of about 600 miles per hour ($\approx$ 965.60 km/h). If this assumption were correct, the experiment would proceed as in the figure on the right of the picture above. The cannonball ejected from the cannon in the AC direction would be affected by the momentum of the planet's motion in the AB direction and would follow the path in the AD direction. During this time, the planet would rotate axially so that the cannon should be in position B. At this point, the cannonball would start moving downwards. During the downward movement of the cannonball, the cannon should move to position S due to the axial rotation of the planet, and the cannonball would drop to point B. Point B should be at a considerable distance from point S. The average time the cannonball spent in the atmosphere was 28 seconds: 14 seconds it went up and 14 seconds it went down.

Multiplying this time by the Earth's assumed rotational speed results in a distance of 8,400 feet ($\approx$ 2,560 m = 2.560 km) which means that the cannonball would have fallen about 8,400 feet ($\approx$ 2,560 m = 2.560 km) from the cannon. However, the cannonball always fell only a few inches (1 inch $\approx$ 2.54 cm) from the cannon's muzzle. The result of this experiment shows conclusively that the planet has no rotational motion and is stationary. Astronomers explain the above experiment as follows: Gravity causes the planet's atmosphere to be attracted to the planet so that the atmosphere moves synchronously with the planet's axial rotational motion. The atmosphere pulls the cannonball along with it in its rotational motion so that the cannonball stays virtually stationary, both when propelling it into the air and when falling, roughly in the direction of the cannon's barrel. The assumption that a heavy cannonball is pulled by the planetary atmosphere for a distance of about 2,560 km and with such a high synchronicity that it falls in the immediate vicinity of the cannon and sometimes even into the cannon's mouth is as absurd as all the other assumptions in the convex-heliocentric theory.

When firing the cannon in the east and west directions, the distance between the cannon and the place where the cannonball fell is always about the same. If the planet were spinning from west to east at an equatorial speed of about 600 miles per hour ($\approx$ 965.60 km/h), the cannonball fired in the east direction, i.e. in the direction of the planet's assumed rotation, would have to fall at a considerably greater distance from the cannon than the cannonball fired in the west direction, i.e. in the opposite direction to the planet's assumed axial motion. In fact, regardless of the direction in which the cannon is fired, i.e. whether it is fired east, west, south or north, the cannonball always falls at approximately equal distances. Astronomers have no explanation for this.

During the Crimean War, between 1853 and 1856, the connection between artillery fire and the rotation of the planet was hotly debated by military men, scientists, philosophers and politicians. Here is what the British Prime Minister, Lord Palmerston, wrote to the Minister of War, Lord Panmure, on 20.12.1857: „This is a research which will be important and at the same time easy to make. The question is whether the axial rotation of the planet has any effect on the flight curve of the cannonball. It is assumed that it does have this effect and that is why the cannonball flies pushed by the explosive dust in a straight direction from the mouth of the cannon, and why the cannonball does not follow the rotation of the planet in the same way as when it is at rest on the surface of the earth. If this is so, the cannonball fired in a southerly direction, i.e. directly south or north, would probably deflect west of the target object because during flight that object would move east somewhat faster than the fired cannonball. Testing can easily be done in any place where there is an open circle with a radius of one mile or more. A cannon can be

VIII. Tellus Plana

placed in the center of this circle and detonated alternately north, south, east and west with equal charges to certify that at each firing the cannonball flies the same distance."

There have been several experiments of this kind which have shown that projectiles fired in different directions fly approximately equal distances. This shows that the planet is not moving.

The fastest passenger planes in human history were the Concorde and the Tupolev TU-144. Concorde reached Mach 2.2, which was 2.2 times faster than the speed of sound. Tupolev TU-144 reached Mach 2.0. These passenger planes have been retired and no commercial passenger plane today (2017) reaches the speed of sound. The convex-heliocentric theory assumes that the planet and its atmosphere would spin from west to east so that a point on the planet's surface at the equator would have an average speed of 1,675 km/h, which is higher than the speed of sound because the speed of sound is 1,224 km/h. As we can see, according to this theory, which has never been experimentally proven, the planet would spin from west to east at a speed faster than the speed of sound. If this assumption were correct, no commercial passenger plane in this world could fly from west to east because, in this case, flying east means flying in the direction of the planet's rotation, and if you fly at a speed slower than the planet's rotational speed you will never be able to move forward. However, commercial passenger planes fly east at less than the speed of sound and reach their destination.

Helicopter takeoff: The convex-heliocentric theory holds that the planet is a convex-shaped object rotating around its axis at a tangential speed of about 1,675 km/h.

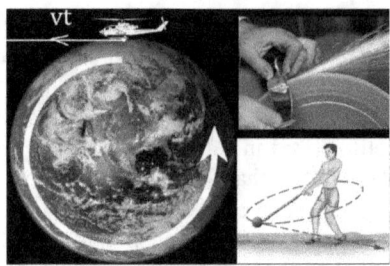

Helicopter lifting power is the force that lifts the helicopter off a surface. This force is a vertical vector that has the sense oriented upwards. The helicopter weight is the force that pulls the helicopter to the ground. This force is a vertical vector that has the sense directed downwards. When a helicopter takes off, the vector of the helicopter's lifting power would be equal in modulus and opposite in sense to the vector of the helicopter's weight. Consequently, at takeoff, the resultant of the two vectors would be zero. A helicopter taking off from such a planet would be propelled at a speed of 1,675 km/h in the direction and tangential velocity vt of the convex planet, in exactly the same way that the iron filings detached from a grinder are projected in the direction of the tangential velocity of the grinder. This phenomenon should occur every time a body is detached from the ground. Such a phenomenon has never been observed on this planet. The planet is stationary.

THE EARTH IS INFINITE: The Earth is an infinite area. There are many planets in this area. The number of planets keeps increasing and decreasing. Above the Earth are the atmospheric layers that together form the Earth's atmosphere. Above the Earth's atmosphere lies the gaseous atmosphere of outer space.

VIII. Tellus Plana

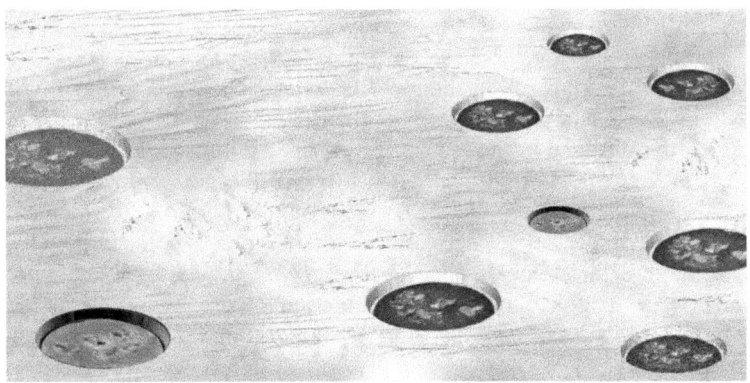

An interesting and astonishing feature of the Infinite Earth is that when you look at the sky you are also looking at the Earth. If you look above the horizon and if you could remove the Earth's atmosphere and the gaseous atmosphere of outer space, you could see the far territories of the Infinite Earth. In the following lines I explain this assertion. Imagine you are on the ground in an area where there is nothing in front of you to block the horizon! This area may be on a vast plain, in a desert, or on the shore of a large body of water such as a sea or ocean. From you to the horizon you will see the land as plain if you are on a plain, as sand if you are in a desert, or as water if you are on the shore of a wide expanse of water. Above the horizon you will see the sky. The sky is nothing more than the earth's atmosphere and the gaseous atmosphere of outer space. The approximate distance to the horizon can be calculated with the following formula:

$$d \approx 3{,}57\sqrt{h}$$

d = distance in km to the horizon.
h = height from ground to eye measured in metres from sea level

(info: sea level is considered as zero level). For an observer standing at sea level, whose eyes are 1.70 m above sea level, the distance to the horizon is about 4.7 km. For an observer at the top of Mount Everest, so for an observer standing about 8.84 km above sea level, the distance to the horizon is about 336 km. Imagine being at sea level and looking at a point just above the horizon! When you climb higher, this point will be on the ground. Therefore, the moment you looked at the point above the horizon, so at a point in the sky, you were actually looking in the direction of the earth. To be more precise, you looked both in the direction of the earth and in the direction of the sky because from the horizon line, the earth merges with the sky.

Consequently, whichever direction you look, even if you are looking at the sky, that direction leads to both sky and earth. To understand this even better I'll give you another example. Imagine that you are back to the place where you can see the horizon without something obstructing your horizon! Look down at your feet and then look at the ground further and further away from your feet in a direction perpendicular to the horizon line until you touch the horizon line with your eyes! You will notice that to get to look at the horizon line, you will have to raise your gaze higher and higher. In other words, the ground between you and the horizon appears to be sloping upwards from your feet towards the horizon. Beyond the horizon, the Earth continues its ascent, but can no longer be seen due to its obstruction by the Earth's atmosphere and the gaseous atmosphere of outer space. It is a characteristic of the Infinite Earth that it appears to be oblique and ascending into the distance.

If you look around your location for a short distance, the Earth appears flat. This appearance coincides with reality. If you look around for a long distance and towards the horizon, the Earth appears to be tilted upwards towards the horizon. Because of these two aspects, you have the impression that you are in a concave shape. This experience is harder to do on the ground because it is not so

VIII. Tellus Plana

easy to find a place where the area from you to the horizon and around you does not have something obstructing the horizon. However, you can observe this phenomenon very well if you are on the open seas or oceans. You can also see it very well if you are in the air, for example in a hot air balloon. Everyone who has travelled in a hot air balloon has had the sensation that the ground below them is concave because the Earth appears to be tilted upwards in the distance between the observer and the horizon. Here are some accounts from balloonists.

Here is what the London Journal newspaper wrote in its edition of 18.07.1857: „The main characteristic feature of an image from a balloon, at a considerable altitude, was the attitude of the horizon which remains practically at eye level at an altitude of two miles (info: 2 miles ≈ 3.22 km) making the Earth appear concave instead of convex and receding during rapid ascent while the horizon and balloon appear stationary."

In his book The Great World of London, Henry Mayhew (25.11.1812 - 25.07.1887) stated: „Another curious effect of the aerial ascent was that the earth, when we were at our highest altitude, appeared unmistakably concave, appearing as a gigantic dark vessel rather than a convex sphere as we would naturally expect to see it. The horizon always seemed to be at our eye level and seemed to rise until finally the altitude of the circular contour line of the field of view became so obvious that the earth took on the abnormal appearance of being more of a concave entity than a convex one."

In the edition of the British weekly newspaper The Leisure Hour of 21.05.1864, the English meteorologist, astronaut and astronomer James Glaisher (07.04.1809 - 07.02. 1864) wrote: „The plane of the earth gives another illusion to the air traveller, to whom it appears as a concave surface, and who looks upon the horizon as an unbroken circle rising to the eye of the observer at whatever height he may be. The cavity of the concave hemisphere is like the frame of an overturned watch glass. The blue atmosphere above encloses it like an analogous inverted hemisphere."

IX. The Sun

The convex-heliocentric theory holds that the Sun is a nearly perfect sphere made of hot plasma. This is not true at all. Look very carefully at the sunrise and sunset! Notice that the Sun is composed of a solar disc, a solar halo and distinct rays of light! The solar disc is a disc of a colour very close to white. The solar aura looks like a washer and is yellowish in colour. The distinct rays of light coming from the Sun appear to come from the solar aura, but they come from the solar disc. The Sun produces two types of light on the planet, uniform sunlight and distinct rays of light.

Eternal Fire: Above our planet is the Earth's atmosphere. Earth's atmosphere is illuminated due to the phenomenon of light scattering, which will be explained in the Moon chapter. Sunlight is projected onto air particles. Air particles scatter light in all directions. In this way, the Earth's atmosphere is uniformly illuminated. Above the Earth's atmosphere is the cosmic atmosphere. The air becomes thinner as the altitude increases, making the cosmic atmosphere contain extremely thin air. As a result, the air of the cosmic atmosphere cannot scatter much light. This is why the cosmic atmosphere is very dark. The solar ray bundle in the picture has been drawn on the cosmic atmosphere to make it easier to understand how the Sun works.

However, the cosmic atmosphere is so dark that this bundle of rays is practically invisible. Above the cosmic atmosphere is the Eternal Fire. The cosmic atmosphere is made up of clouds of gas. On the Cosmic Atmosphere there is space that has a very high degree of translucency. Through this space penetrates the light and heat of the Eternal Fire. The Sun is in fact the heat and light of the Eternal Fire penetrating through a space with a high degree of translucency. This translucent space I will call the solar structure. Distinct rays of light emanate from the solar disc and are not visible in the space occupied by the cosmic atmosphere, but only in that occupied by the earth's atmosphere. The distinct rays of light occur because there are small spaces on the solar structure that are more translucent than the rest of the space on the solar structure. These rays of light are not the same as those coming from the Sun when the Sun's light penetrates through the clouds on Earth's atmosphere, although they are very similar. The Sun's rays in the image below come from sunlight penetrating through clouds in the Earth's atmosphere.

IX. The Sun

Eternal Fire is fuelled by hydrogen which comes from the Earth. Between 1970 and 1989, the former USSR produced a borehole, as part of a scientific project, 12,262 metres deep on the Kola Peninsula. During the drilling at Kola, scientists found that the mud extracted from the hole was full of hydrogen. As a result, hydrogen is coming out of the ground.

Hydrogen is the lightest chemical element and is easily flammable. It can therefore rise continuously from the ground through the Earth's atmospheric layers and through the cosmic atmosphere into Eternal Fire space where it ignites. This is how the Eternal Fire is fuelled. The Eternal Fire is also fed with oxygen. Oxygen occurs in the process of photosynthesis carried out by phytoplankton and plants. It is a combustion agent. Hydrogen plays the role of fuel for the Eternal Fire and oxygen sustains the burning of the Eternal Fire.

Day-Night Cycle: In the following I will explain how the day-night cycle appears and works on the planet.

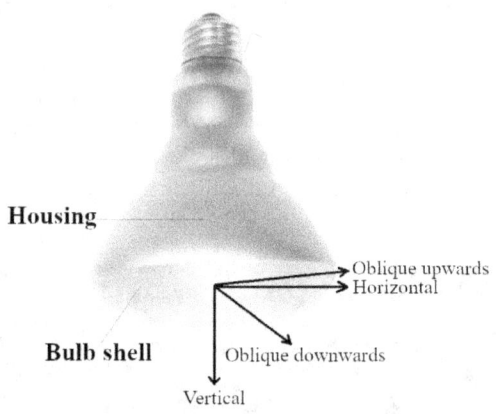

The solar structure through which the light and heat of the Eternal Fire enters, seen from a distance, looks roughly like the bulb shell of the light bulb in the picture above, so it looks a bit like a cap cut out of a sphere. Because this structure looks like the bulb shell of a light bulb, the Sun projects its light locally and much like a flashlight.

IX. The Sun

Just as a flashlight produces a spot of light on the illuminated surface, so the Sun produces a spot of light on the flat planet.

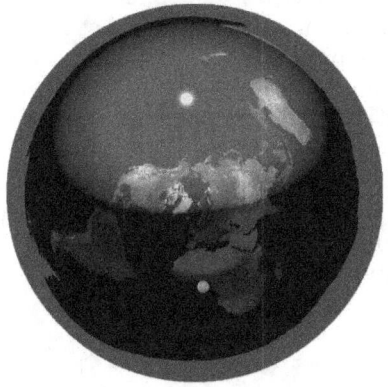

Through the solar structure, light is scattered vertically, in an oblique downward direction, horizontally and in an oblique upward direction. Sunlight scattering obliquely upwards illuminates the space around the solar structure, producing the solar aura. The solar structure contains more translucent and less translucent layers. Light passing through the less translucent layers produces uniform light. Light passing through the more translucent layers produces distinct rays of light that are brighter than uniform sunlight. The Sun's path is always a circle. A circle is always 360 degrees. A solar day lasts 24 hours. Therefore, the Sun rotates 360/24 = 15 degrees per hour. On the surface where the Sun's spot of light appears, it is day, and on the remaining unlit surface it is night. The cosmic atmosphere rotates which causes the Sun to rotate above the planet. This is how the day-night cycle occurs.

Seasons: There is no hemisphere on the planet, only hemiplane. The hemiplane can also be called an hemidisc. The northern hemiplane is the region from the North Pole to the equator, and the southern hemiplane is the planetary area between the equator and the Antarctic Circle. The following picture shows the movement of the Sun. The Sun moves in a clockwise spiral up and down above the flat planet on the surface of an imaginary truncated cone with the top base above the Tropic of Cancer and the bottom base above the Tropic of Capricorn.

The radius of the circle forming the Tropic of Cancer is smaller than the radius of the circle forming the Tropic of Capricorn. The orbital-tangential velocity is directly proportional to the radius which means that the Sun is continuously changing its orbital-tangential velocity. The closer the Sun is to the Earth, the faster its orbital-tangential Velocity will be.

IX. The Sun

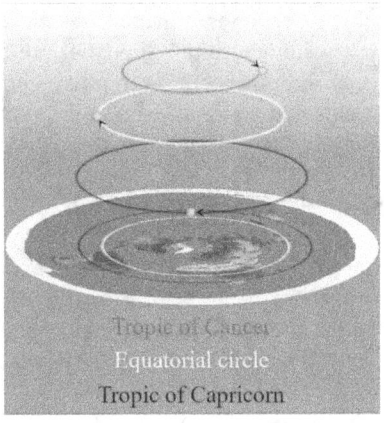

The Sun's orbital-angular velocity always remains the same. The Sun's path is a circle. The circle is 360 degrees. The Sun makes one complete rotation in 24 hours. The Sun's orbital-angular velocity is found by dividing 360 degrees by 24 hours. Therefore, the Sun's orbital-angular velocity is 15 degrees per hour. Expressed in radians, the Sun travels 2¶ in 24 hours.

Therefore, the Sun's orbital-angular velocity expressed in radians is 2¶/24 hours = 0.26 radians per hour. Two principles of perspective have been ignored in the above explanation. These two principles are as follows: a. Shrinking trajectories with increasing altitude at which these trajectories lie. b. Curving the atmosphere and turning it into an apparent dome.

In the following I want to apply the first principle of perspective. In this case, the radius of the orbit in which the Sun is moving appears to decrease relative to the actual radius of the same orbit as the altitude at which the Sun is located increases. Since the radius of the orbit appears to change with the altitude at which the Sun is located, the Sun's orbital-tangential velocity also appears to change with respect to the actual orbital-tangential velocity, but the Sun's orbital-angular velocity remains constant. As the Sun ascends the truncated cone, the Sun's orbital-tangential velocity appears to become less and less than the actual orbital-tangential velocity, and as it descends the truncated cone, the Sun's orbital-tangential velocity appears to become greater and greater than the previous orbital-tangential velocity. From this, we can see that the perspective changes the Sun's orbital-tangential velocity. The change in this velocity is, however, only an appearance produced by the perspective. In the following picture the movement of the Sun is shown taking into account the second principle of perspective. The ellipse representing the flat earth does not represent the whole planet, but only the portion of the earth covered by the apparent dome.

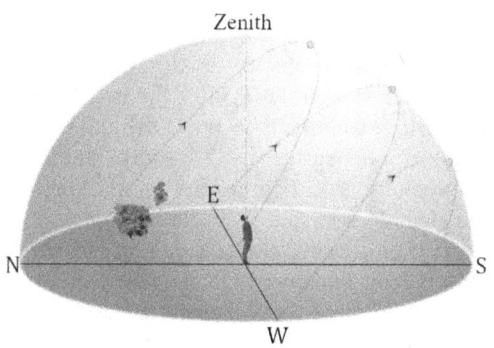

The portion of land is shown to be flat, so it is shown as it really is. The perspective, however, curves it so that it becomes an inverted dome, so it becomes a concavity. The inverted dome appearance of the earth's surface can be seen by an observer standing in the middle of the ocean and looking around the horizon. The horizon line will be a circle demarcating a horizontal surface. The lines below the horizon will be smaller and smaller circles at lower altitudes and will mark horizontal surfaces.

IX. The Sun

Consequently, all these lines will describe an upside-down dome. The inverted dome appearance of the Earth's surface can also be seen from the aerial perspective of a passenger in an air balloon at an altitude of about 10 km. In the image, the atmosphere is curved by the perspective which creates an apparent dome. The height of the apparent dome varies. When the Sun is at aphelion, so when it is at its highest altitude, the height of the apparent dome is maximum. When the Sun is at perihelion, i.e. when it is at its lowest altitude, the height of the apparent dome is the minimum. The Sun rotates on this apparent dome as in the picture.

The trajectories in the image are the Sun's trajectories as seen from locations on the northern hemiplane. The path on the left is the path of the Sun during the solstice on 20 June and 21 June respectively. During this period, the Sun rises in the north-east and sets in the north-west. The middle path represents the path of the Sun during the equinoxes on 20 March and 22 September and 23 September respectively. During this period, the Sun rises in the east and sets in the west. The path on the right is the path of the Sun during the solstice on 21 December and 22 December respectively. During this time, the Sun rises in the south-east and sets in the south-west. There are three points to note here: a. The visual sense cannot distinguish the growth of the apparent dome. In other words, to the visual sense the apparent dome is always the same height. b. Because the height of the apparent dome appears to be the same, the radius of the orbit in which the Sun moves appears to be the same throughout the year. c. On the northern hemiplane, during the June solstice, there is sunlight for a longer period of time than during the December solstice. In other words, the Sun stays in the sky longer during the June solstice than during the December solstice.

These three aspects give the impression that the Sun would change its orbit-angular speed so that during the June solstice it would be slower than during the December solstice. The Sun is over the Tropic of Cancer between 21 June and 22 July. While the Sun is over the Tropic of Cancer, the days are longer and warmer on the northern hemiplane. During this time, for those near the Sun's centre of rotation, i.e. near the North Pole, the Sun is visible 24 hours a day. The Sun is above the Equatorial Circle during the Spring and Autumn Equinoxes. It is over the Tropic of Capricorn between 21 December and 22 January.

While the Sun is over the Tropic of Capricorn, the days are shorter and cooler on the northern hemiplane. During this time, the Sun moves at its fastest speed because it travels a greater distance in 24 hours. This is why the days are shorter during winter in the northern hemiplane. This movement of the Sun explains very well why in the equatorial regions there is summer for most of the year, while in the northern and southern latitudes there are distinct seasons with harsh winters. Because the Sun's path on the Tropic of Capricorn is the lowest path, the winter Sun is at its lowest position in the sky.

Arctic Midnight Sun: The Midnight Sun in the Arctic proves that the Sun revolves above our planet. There is no Midnight Sun in Antarctica. During the Arctic summer, which takes place from 22-25 June, starting at 66 degrees 33 minutes north latitude and from a high altitude, you can observe a phenomenon known as the Midnight Sun, which consists of a three-day observation of the Sun. The Midnight Sun rises on 22 June and remains in the sky for 3 days (= 72 hours). It slowly rises and sets every 12 hours, resulting in three twilights and three bright sunrises without the Sun disappearing below the horizon. The convex-heliocentric theory holds that the planet would revolve around its axis and around the Sun. This former theory can be called Earth without axial precession. In this theory, the Midnight Sun in the Arctic would not be visible from a northern latitude of less than 89 degrees because you would have to look many kilometres through the Earth.

IX. The Sun

Terra without axial precession

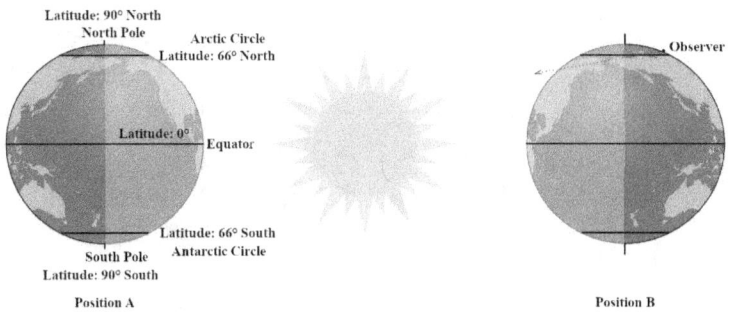

Position A   Position B

But it is visible even from a latitude of 66 degrees north. In position B, in the image above, the left side of the planet is illuminated by the Sun, so it is daytime, while the right side of the planet is not illuminated by the Sun, so it is night. An observer is inside the Arctic Circle. He is on the right side of the planet inside the Arctic Circle and so it is night. This observer, in order to see the Sun at Midnight in the Arctic, would have to look many kilometres through the Earth. In the case of Earth theory without axial precession, the phenomenon called the Arctic Midnight Sun cannot exist. Consequently, this theory is not in line with reality and is therefore false. In order for the astronomers' theory to stand, these astronomers have modified their theory in order to allow for the fact that the Midnight Sun in the Arctic is visible from 66 degrees north latitude. This gave rise to a theory we can call the Earth with axial precession. This theory states that the planet's axis would be tilted relative to another axis at an angle of 23.5 degrees and that it would rotate around two axes. Under this theory, the planet would rotate about its axis while its axis would rotate about the other axis. This theory assumes that the planet's axis would make a complete rotation around the other axis in about 25,772 years. In this case, there are two basic situations.

Terra with axial precession
Case 1

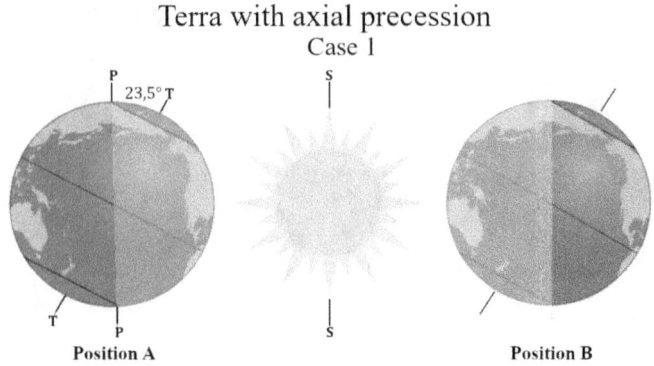

Position A   Position B

The image above shows the first situation. The T-T axis is the axis of the planet. The P-P axis is the axis around which the T-T axis revolves. The S-S axis is the axis of the Sun. This situation arises when the P-P axis, the T-T axis and the S-S axis lie in the same plane. This situation would occur only once during a precession cycle lasting about 25,772 years. In this situation two positions occur, namely the A position and the B position. In position A, the Midnight Sun could be seen all around the Arctic Circle, so even from latitude 66 degrees, but it could be seen not only in the period 22-25 June, but over a very long period of time. At position B, the Arctic Midnight Sun would not be seen at all, but instead the phenomenon that can be called the Antarctic Midnight Sun should appear. But there is no Midnight Sun in Antarctica.

The following picture shows the second situation. This situation would occur most of the time except when situation 1 occurs. In situation 2, notice that the angle between the T-T axis and the P-P axis appears to be smaller than in situation 1 because the T-T axis is tilted backwards. In situation 2, the Midnight Sun in the Arctic would not be seen all around the Arctic circle formed by the circumference of 66 degrees latitude, and should also be seen in Antarctica. In

IX. The Sun

Antarctica, however, there is no Midnight Sun. The convex-heliocentric theory does not coincide with reality and is therefore false. Only on the flat planet can the Midnight Sun appear in the Arctic. On a flat planet, the Sun always sits above the planet. During the summer, i.e. 22-25 June, it comes very close to the North Pole. This is the reason for the phenomenon known as the Arctic Midnight Sun.

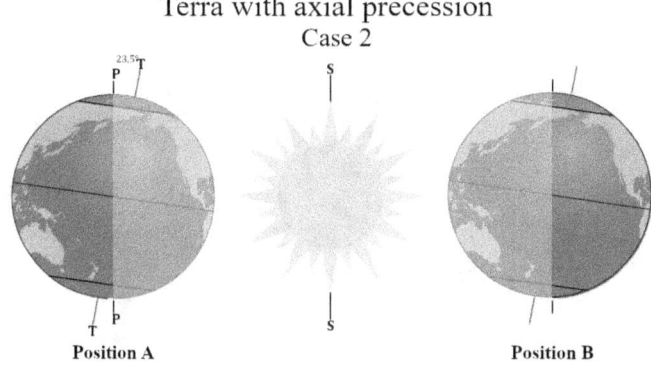

Terra with axial precession
Case 2

Position A   Position B

X. The Moon

The Moon has always been the most mysterious celestial entity. It seems to change its shape, size, colour, brightness and position in the sky. There is a Supermoon and a Micromoon. The Supermoon is the Moon in its Full Moon phase, which occurs when the Moon is at perigee, i.e. when the Moon is at its lowest altitude. The micromoon is the Moon in the full Moon phase that occurs when the Moon is at apogee, i.e. when the Moon is at its highest altitude. The Moon can be different shades of grey, blue, yellow, orange or red. Its brightness varies. For example, the brightness of the Supermoon is higher than the brightness of the Micromoon. The Moon changes its position in the sky by arising at the horizon and rising in the sky and then setting behind the horizon.

The Moon's surface has a very interesting structure. There are special effects produced by the Moon. One such effect is when the Moon's surface is composed of a very bright surface that is shaped like a lunula and a less bright surface that appears to be elliptical. The lunula is the shape of the curved part of a sickle. The moon has a cyclic motion. This movement is called the lunar cycle. The lunar cycle takes about 30 days. The period of about 30 days is called a calendar month. Because the moon has a cyclical movement, its movement can be used to measure time and so calendars can be made based on this movement. These calendars are called lunar calendars because they are based on the movement of the moon. Even in solar calendars, the movement of the moon is taken into account because in these calendars the year is divided into 12 calendar months, each of which has a number of days equal to the number of days of the lunar cycle. The moon is an extremely complex phenomenon. This phenomenon is so complex that contemporary astronomers have been unable to provide any scientific and accurate information about the Moon, despite having supergiant funding and the most modern equipment in the world. Even in the scriptures of ancient sages there is no scientific and thorough information about the Moon. The Moon is such a mysterious entity that all the cultures of the world have elevated it to the status of a deity. It has remained a mystery to this day. In this chapter, I will reveal this mystery.

Definitions and Explanations: Before explaining the Moon, some preliminary information is needed. In this subchapter, I give you the definitions of the basic concepts and the preliminary information you need to understand the Moon.

Image resolution: This section is about the resolution of light entities in the sky, not the resolution of images from photography or filming. Image resolution is the amount of detail of a luminous nature that comes from the surface of an illuminated entity or a luminous projection on the atmosphere, such as the Moon. The higher the brightness of the entity, the greater the amount of light detail coming from that the entity and therefore the higher the resolution of that is. The lower the brightness of the entity, the lower the amount of detail of a luminous nature coming from the entity and therefore the lower the resolution of the entity. Optical instruments, such as telescopes, and thus instruments that can magnify distant features in order to observe details, can only magnify that feature to the maximum resolution of that feature. When the maximum

X. The Moon

resolution is reached, the resolution can be said to have been consumed. If after the resolution has been consumed the entity is further enlarged, then no new details will be seen, but the image will become increasingly distorted. Magnification of bright entities in the sky using optical instruments can be simulated on the computer. Images 1, 2 and 3 are an example of such a simulation. Image 1 shows the Moon in the night sky. At the bottom of the moon there is an area where there is an almost white circle surrounded by a darker ring. This area we have magnified by a factor of 2, i.e. it has been magnified twice. In image 2 you can see this area magnified by a factor of 2. As you can see, new details appear in this area and the image is clear. At the top of the circle you can see small rays that appear to come from this circle. These small rays are an example of the new details that appear by magnifying image 1.

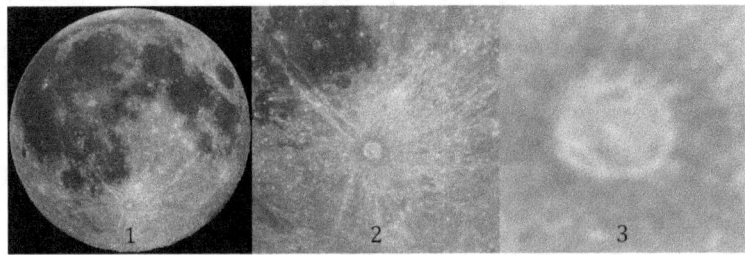

That area was increased by a factor of 20. In image 3 you can see this area magnified by a factor of 20. As you can see, no new details appear in this area and the image is blurred. We can say that the resolution of the image has been consumed because no new details appear on the image. If we enlarge this area even more, no new details will appear, but the image will become even more distorted. Apart from the Sun, the Moon is the brightest entity in the sky. A modern camera can have an optical zoom of 125x and a digital zoom of 250x. Such a camera can easily consume the Moon's resolution.

The largest giant telescopes in the world are, in order of size, the following: the Great Telescope of the Canary Islands on the island of San Miguel de la Palma in Spain, the W. M. Keck Astronomical Observatory consisting of the Keck 1 and Keck 2 telescopes located near the summit of an inactive volcano called Mauna Kea in the state of Hawaii, USA, the Large South African Telescope located in the town of Sutherland in South Africa and the Large Binocular Telescope located on Mount Graham in Arizona, USA. There are several giant telescopes in the world, but none of them can take more detailed images than those produced by a modern camera because of the aforementioned. It is possible that these telescopes are used for purposes other than those officially stated.

Atmospheric pressure: Atmospheric pressure, also called barometric pressure, is the pressure inside the Earth's atmosphere. As altitude increases, atmospheric pressure decreases. In other words, the atmospheric pressure at a high altitude is lower than the atmospheric pressure at a low altitude. Decreasing atmospheric pressure leads to decreasing air density. In other words, there is less air at higher altitudes than at lower altitudes. This is why the Earth's atmosphere has a higher density than the cosmic atmosphere. The cosmic atmosphere has an extremely low density and is composed of light gases such as hydrogen, helium, nitrogen, oxygen and carbon dioxide.

Orbital-tangential and orbital-angular velocity: Both orbital-tangential and orbital-angular velocities have been described in the chapter entitled Successive Explanations. These types of velocity are important to this chapter because they have to do with the Moon.

X. The Moon

Light reflection:

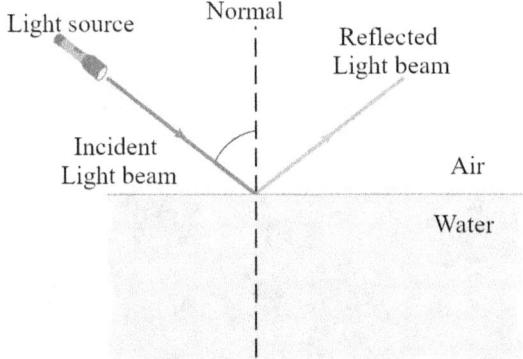

The picture above shows how reflection works in principle. Light reflection is the phenomenon of changing the direction of propagation of a ray of light by bouncing the ray when it hits the surface of a medium other than the one from which it came. The ray of light coming from the light source is called the incident ray. The surface which the incident ray strikes and bounces off is called the reflecting surface. In this case, the reflecting surface is the surface of the water. The line perpendicular to the reflecting surface is called the normal to the reflecting surface. The angle that the incident ray makes with the normal to the reflecting surface is called the angle of incidence. The angle that the reflected ray makes with the normal is called the angle of reflection. The medium in which the reflection takes place is called the reflecting medium. In this case, the reflecting medium is air. The incident ray travels towards the reflecting surface, touches this surface and changes its direction of propagation by bouncing away from this surface. The law of reflection states that the angle of incidence is equal to the angle of reflection. Reflection can occur in the same medium. The mirror is one of the instruments used which is based on the phenomenon of light reflection.

Refraction of light:

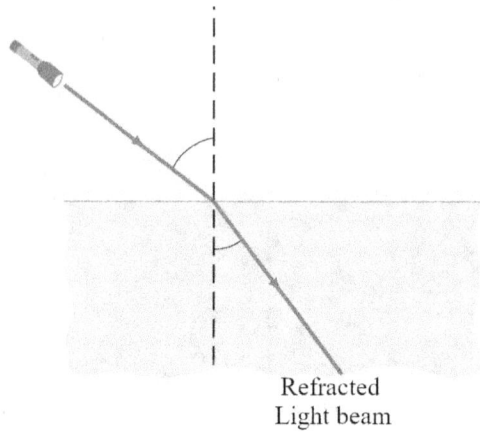

Refracted
Light beam

The picture above shows how refraction works in principle. Refraction of light is the phenomenon of changing the direction of propagation of a ray of light by reflecting the ray as it passes through the surface separating two different media. The ray of light coming from the light source is called the incident ray. The surface that is touched by the incident ray and through which this ray passes from one medium to the other medium is called the refracting surface. In this case, the refracting surface is the surface of the water.

The line perpendicular to the refracting surface is called the normal to the refracting surface. The angle that the incident ray makes with the normal to the refracting surface is called the angle of incidence. The angle that the refracted ray makes with the normal is called the angle of refraction. The medium in which refraction takes place is called the refracting medium. In this case, the refracting

X. The Moon

medium is water. The incident ray travels towards the refracting surface, touches this surface and changes its direction of propagation by deflecting away from this surface. The law of refraction states that the angle of incidence is not always equal to the angle of refraction. Refraction requires two different media. The lens is one of the instruments used which is based on the phenomenon of refraction of light. In the next image on the left, the Sun is at twilight near the horizon, causing the solar disc to produce an elongated streak of light on the surface.

The surface of the water appears to slope upwards towards the horizon. This tilting effect of the water surface is an appearance due to perspective. The plane of the solar disk appears to be vertical. The verticality of the solar disc is an appearance due to perspective. Consequently, the apparent angle between the surface of the water and the solar disc is slightly more than 90 degrees. In the image on the right there is a puddle in which the reflection of the solar disc can be seen. The puddle is made of water. Above the puddle is air. In the sky, so above the puddle, is the solar disc. Air is a less dense medium than water. The reflection of the solar disc takes place exactly on the surface of the water, i.e. on the upper surface of the denser medium. We can say that the solar disc is projected onto the water. With the following picture we can see more about the phenomenon of light reflection.

We notice that the sunlight enters through the window and is projected onto the floor. The floor surface on which the sunlight falls reflects this light in the elongated shape of the window through which it enters. This light effect is based on the same principle as the formation of an elongated streak of light by the sun disc on water when the setting sun is near the horizon. The light entering through the window is reflected on the floor in the approximate shape of the window through which it enters. We can say that the light shape of the window is projected onto the floor. This light effect is based on the same principle as the reflection of the sun's disc in the water of a puddle when the sun is over the puddle. This is also the principle behind the formation of the Moon. The Moon is the reflection of the

X. The Moon

Sun's disc that is formed when sunlight passes from a less dense to a denser medium. In this case, the less dense medium is the cosmic atmosphere and the denser medium is the Earth's atmosphere. The Moon is the projection of the solar disc onto the Earth's atmosphere.

Light dispersion: The picture shows the working principle of dispersion. Light dispersion is the phenomenon whereby a beam of white light is split by refraction into coloured beams of light as it passes through a particular medium. The medium that disperses light is called the dispersive medium. The coloured beams of light have colours that make up the spectrum of white light. These colours are red, orange, yellow, green, blue, indigo and violet.

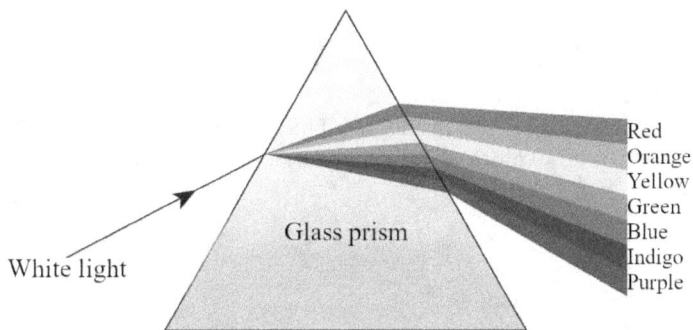

Light diffusion: The following picture shows the working principle of the diffusion. Light diffusion is the phenomenon whereby a ray of light strikes a particle and is deflected from its direction by that particle in several directions. In colloquial language, diffusion may also be called scattering. Air molecules diffuse a large amount of blue beams and a very small amount of beams of other colours.

This is why the sky is blue in most cases. Because air becomes thinner with increasing altitude, there are very few particles in the cosmic atmosphere that could diffuse light and brighten the cosmic atmosphere. This makes the cosmic atmosphere a very dark area.

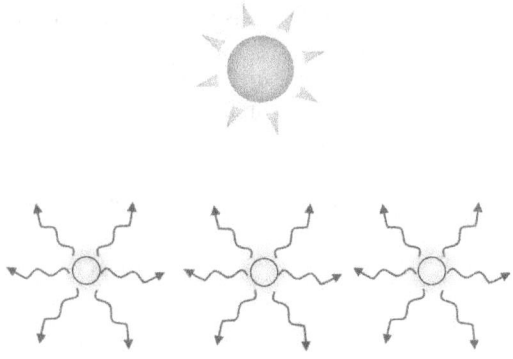

In the next picture is the inside of a building. With the help of this image we can explore the phenomenon of reflection and light diffusion.

X. The Moon

Sunlight penetrates through the windows of the building and produces two types of light. These two types of light can be called uniform light and light in the form of distinct rays of light. Uniform light in the building is due to the phenomenon of light reflection. Light entering through the windows of the building is reflected by the walls and floor. In other words, the walls and floor reflect light entering through the windows of the building many times, creating uniform light inside the building. There is a lot of dust in this building. Distinct rays of light are formed due to the phenomenon of light diffusion caused by dust particles in the building. Dust particles are larger than air particles. This is why dust particles can diffuse white light, i.e. they can diffuse the entire light spectrum. Light that encounters dust particles is diffused by these particles, so it is scattered in many directions. Light diffused by dust particles is white. Because it is diffused in many directions by all the dust particles in that area, the light in that area becomes bright. The brightness of this light is because it is brighter than uniform light. Because diffused light is brighter than reflected light, diffused light appears as distinct rays of light.

Aquatic vision: In the following image is a photo taken from an aquatic perspective. In this picture you can see the solar disc, the cosmic atmosphere, the Earth's atmosphere and the sun's rays. The solar disc is the bright part from which the sun's rays emerge and is roughly circular in shape. Notice that around the solar disc there is a less illuminated ring! This ring is actually the very faint illuminated surface of the cosmic atmosphere.

X. The Moon

As mentioned above, the cosmic atmosphere is very poorly lit because it is very rarefied, so air particles cannot scatter much light around them. This is why a dark, roughly ring-shaped surface appears around the solar disc. This ring is penetrated by the sun's rays that appear in the Earth's atmosphere. In the following picture there are two circles and four white lines. The white lines are the demarcation lines. The small circle surrounds the solar disc. Between the small and large circles is a surface in which both the cosmic atmosphere and solar rays are mirrored. Between the great circle and the demarcation lines is a surface that mirrors the Earth's atmosphere and the sun's rays. The boundary lines indicate the edges of the total light surface. This surface is roughly round in shape. The camera lens was not pointed in a direction perfectly perpendicular to the plane of the solar disc and passing through the centre of the solar disc, but in an inclined direction. Therefore, the centre of the solar disc is not in the same position as the centre of the great circle.

Tellus Plana

The light from the solar disc passes through the cosmic and terrestrial atmosphere and is projected onto the water. The Earth's atmosphere is made up of air. Air is a medium that has a lower density than water. Consequently, the solar disc projects onto water, i.e. it projects onto the upper surface of the higher density medium. The light from the solar disc passes further through the water which makes it visible from an underwater perspective. As does light from the solar disc, so does light from the Earth's atmosphere. The Moon is formed according to the principle that the entire surface area of light in the image is formed. There is, however, a small difference between the formation of the Moon and the formation of the total light surface. The Moon forms on the surface of the Earth's atmosphere by passing sunlight through the cosmic atmosphere until it reaches the Earth's atmosphere. The total light surface in the picture is formed on the surface of water by passing sunlight through the cosmic atmosphere and through the Earth's atmosphere until it reaches the surface of the water. Therefore, only the solar disk, the cosmic atmosphere, some of

X. The Moon

the light from the Earth's atmosphere and the sun's rays will be reflected on the Moon. The Moon has a certain thickness which is less than the height of the Earth's atmosphere. This is why only some of the light from the Earth's atmosphere can be seen on the Moon.

Image 1 shows the photo taken from an underwater perspective. The total area of light appears to be convex. In the area at the edge of this surface there are elements in the form of ripples. Image 2 shows the Moon. The Moon's surface also appears to be convex and has similar elements at the bottom edge. These features are very small, so you can see them better if you enlarge a high-resolution image of the Moon on your computer. In other words, there are similarities between the total area of light in image 1 and the Moon. These similarities exist because the phenomenon of the formation of the total light surface is similar to the phenomenon of the formation of the Moon. Now I will reveal the first secret. The elements that appear on the total light surface in the form of the solar disc, the ring of the surface of the cosmic atmosphere and the solar rays also appear on the surface of the Moon. On image 3 are 2 circles. The small circle is superimposed on the solar disc. The cosmic atmosphere is mirrored between the small and large circles. From the great circle to the edge of the Moon is the projection of the Earth's atmosphere.

The solar disc is not in the centre of the Moon because of the perspective. I will explain this in the following lines.

X. The Moon

In picture 1 is a photo showing, among other things, a relatively long street with lanterns on the edges. This picture helps us understand two principles of perspective. The first principle is that objects appear smaller and smaller the greater the distance between the observer's eyes and the observed entity. The greater the distance between the observer and the lantern light, the smaller the lantern light appears to be. The second principle is that everything above the observer's eyes seems to descend more and more as the distance between the observer and the observed entity increases. The lantern lights are above the observer's eyes. Therefore, they seem to get lower and lower as the distance between the observer and the observed light increases. In Fig. 2 there is the photograph from Fig. 1 over which two elements have been superimposed: a gymnastic hoop called a hula-hoop and a disc with a white surface. I made this image to get a better picture of what happens to the Sun's disc when viewed through the Moon.

Imagine that the hula-hoop circle is at a short distance from the observer and above the observer's eyes! Imagine that at a great distance from the observer there is a white disc of cardboard, that the diameter of this disc is equal to the diameter of the hula-hoop circle, that the surface of this cardboard is parallel to the surface of the hula-hoop circle and that the altitude of the centres of the two elements is the same! Now apply the two principles of perspective stated above. If you do this you will notice that the cardboard disc will appear shrunken and displaced downwards from the centre of the hula-hoop circle. As a result, the solar disc on the Moon's surface is not in the centre of the Moon, but appears smaller and shifted downwards from the centre of the Moon due to the two principles of perspective.

Apparent dome: The retina is a layer with a maximum thickness of 0.5 mm. This layer is located on the inner side of the eyeball wall and is made up of photosensitive cells. In adult humans, about 72% of the retina is shaped like a sphere. This approximate sphere is about 22 mm in diameter. Because the retina is approximately spherical and the image is projected onto the concave side of the retina, the visual system always produces a concave image. Sometimes, due to the structure inside the circle representing the concavity and the different degrees of brightness of the colours inside the circle, the concavity is perceived to be convexity. In the following illustration there are two images. In image 1 we have represented reality without taking into account the principle of perspective whereby the sky is transformed into an apparent dome. Above the planet is the earth's atmosphere. Above the Earth's atmosphere is the cosmic atmosphere. Above the cosmic atmosphere is the Eternal Fire. Picture 2 shows reality in terms of the principle whereby the sky is transformed into an apparent dome. The edge of the apparent dome is on the cosmic horizon. When the Sun is at aphelion, i.e. when it is at maximum altitude, the apparent dome has a maximum height. When the Sun is at Perihelion, i.e. when it is at minimum altitude, the apparent dome has a minimum height. Consequently, the maximum height of the apparent dome varies. As you can see in the image above, the Moon is always below the Sun.

Here the next question arises: If the Moon is always below the Sun, how is it possible to observe the Moon in the sky in a different position than the Sun? The Moon may appear in the sky in a different

X. The Moon

position from the Sun due to perspective. I will answer this question in the subchapter entitled Moon Motion. In order to explain the Moon I think it is good to review the perspective structure chart. Across the distance between the terrestrial horizon (TH) and the celestial horizon (CH), the N2 and N3 layers of the Earth's atmosphere tilt upwards, which causes it to project into the sky. Consequently, across the distance between TH and CH, the sky consists of two overlapping layers: the real atmosphere and its projection. The first projection of the real atmosphere, i.e. the one between TH and CH, we have called the Primary Projection (PP). When the atmosphere projects onto itself, two types of structures are projected onto the surface of the Moon.

The first type of structure we have called the Black Disk. This structure is the projection of the Black Sun. The second type of structure we have called the Solar Spider. The Solar Spider is made up of the following elements: the solar disc, solar rays from the cosmic atmosphere and solar rays from the Earth's atmosphere. I will explain the Black Disk and the Solar Spider in the chapter entitled Elements of the Moon. Let's take the case where the atmosphere projects onto itself exactly at point TH. In this case, point TH can also be called the projection point. From TH to KH there are numerous projection points. As the cosmic atmosphere rotates, which causes the Sun to rotate, at each projection point the two elements, a Black Disk and a Solar Spider, will be projected onto the Moon. The two elements will appear on the Moon's surface in different sizes. The distance from the observer to the TH depends on the altitude of the observer's eyes. If the observer's eyes are 1.70 m above sea level, the distance from the observer to the TH will be about 4.7 km. If the observer is on Mount Everest and so the observer's eyes are about 8.848 km above sea level, the distance to TH will be about 336 km. Consequently, the TH projection point moves instantaneously more and more towards KH as the altitude at which the observer's eyes are located increases.

When the projection point TH moves, all projection points move instantaneously and simultaneously so that the distance between them always remains the same. Because the distance between the projection points always remains the same, the distances between the elements that make up the image of the Moon remain the same and so the image of the Moon remains approximately the same.

Consequently, the centre of the image together with the inner image of the Moon moves with altitude. Their displacement is very small and therefore difficult to observe.

Astronomers have observed this movement. They call it libration motion. The Moon's libration motion is an apparent motion. It occurs when observing the Moon from different altitudes. The altitude between the observer and the Moon increases or decreases due to the movement of the Moon in the sky.

Cloud Study: In the following picture you can see the clouds in the day sky when lit from above. The top of the cloud has a whitish surface, while the bottom of the cloud has a grey surface. Sunlight penetrates through the cloud at the top of the cloud and then passes through all the layers of the cloud and exits through the bottom surface of the cloud. There are small water droplets in the cloud. These water droplets are much larger than air particles. When sunlight comes into contact with an air particle, that particle diffuses a large amount of the small wavelengths of light. Blue light is light with a short wavelength. Consequently, air particles diffuse a large amount of blue light. This is why the sky is blue. In the cloud, sunlight is diffused by water droplets which are much larger than air particles. Because the water droplets in the cloud are large, they diffuse all the visible wavelengths of the solar spectrum which makes the top of the cloud appear to be whitish. The amount of this sunlight decreases as the light penetrates through more layers of the cloud.

X. The Moon

Eventually, the amount of light that penetrates through the bottom surface of the cloud is much less than the amount of light that has penetrated through the top surface of that cloud. This causes the bottom surface of the cloud to emit less light and so darken in colour to a dark grey. Sunlight passes further through the bottom layer of the cloud, but the brightness of the light coming out of the cloud is equal to the brightness of the atmosphere, so that the light cannot be seen by the observer's eyes. This phenomenon helps to understand how the Moon is formed. When it encounters the Earth's atmosphere, the light from the solar disc is projected onto the Earth's atmosphere and penetrates further through the Earth's atmosphere producing a disc of light that we call the Moon.

The light from the Sun's disc penetrates further through the bottom surface of this disc of light, but from the bottom of the Moon's disc, this light has a brightness equal to that of the atmosphere and so can no longer be seen. The top of the lunar disc is lighter in colour and is what produces the Moon's light. The bottom of the lunar disc is darker in colour and, especially at night, appears to be grey like the bottom of the cloud. This is how the Moon is born.

Formation of the Moon: Above our planet is Earth's atmosphere. Above the Earth's atmosphere is the cosmic atmosphere. Above the cosmic atmosphere is Eternal Fire. The cosmic atmosphere is made up of clouds of gas that can be called cosmic clouds. On the upper surface of the cosmic atmosphere there is a space that is very translucent. Through this space penetrates the light and heat of the Eternal Fire. The sun is in fact the heat and light of the Eternal Fire penetrating through a space with a high degree of translucency. The Eternal Fire is fed by hydrogen coming from the Earth. The burning of the Eternal Fire is sustained by oxygen coming from the Earth. Solar rays pass through the cosmic atmosphere. The cosmic atmosphere is at a very high altitude, so there are no air particles in it. Illumination of the atmosphere occurs due to the diffusion of light caused by air particles. Since there are no air particles in the cosmic atmosphere, it is almost completely unlit.

In other words, the cosmic atmosphere is very dark. The light from the solar disk is truncated cone-shaped. It projects a disc of light onto the surface of the Earth's atmosphere. This disc of light is the Moon.

Because of the truncated shape of the light coming from the Sun's disc, the Moon is somewhat larger in diameter than the Sun. The Sun's rays are projected only onto the Moon's surface. However, the whole of the Earth's atmosphere is illuminated due to the phenomenon of diffusion of light that meets air particles. The Moon's upper surface is whitish, but the colour of its layers gets darker and darker as the altitude decreases so that the Moon's lower layer is darker, usually a greyish colour. Light from the Sun's disc still penetrates through the bottom surface of the Moon's disc, but from the bottom of the Moon's disc to the ground, the light is as bright as the atmosphere and so can no longer be seen. This is how the Moon is born. As you can see in the image above, the Moon is always below the Sun. This leads to the next question: If the Moon is always below the Sun, how is it possible to observe the Moon in the sky in a different position from that of the Sun? The Moon can appear in the sky in a different position from the Sun because of perspective. I will give a detailed answer to this question in the subchapter entitled Moon Motion.

Shape of the Moon: The moon is actually shaped like a disc, but perspective curves the disc into a meniscus-like shape. The meniscus lens is shaped like a watch lens. The following image is an explanatory image to help understand the approximate shape of the Moon. Position A shows the shape of the Moon in profile view. Position B shows the meniscus lens seen from the side. In position C is a watch lens seen from the side. In position D is an image of the Moon. Inside the circle, the Moon has very little curvature, but the curvature of the Moon increases more and more from the circle towards the edges of the Moon. At first glance, the visible surface of the Moon appears to be convex, but it is concave. The Moon's visible surface appears convex because of its structure.

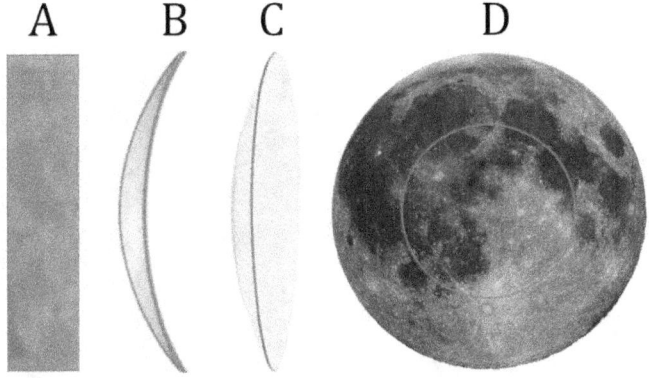

Elements of the Moon: There are two types of elements on the Moon. We call the first type the Solar Spider. The Solar Spider is the projection of the Sun's disc and its rays onto the Moon. The second type we call the Black Disk. The Black Disk is the projection of the Black Sun onto the Moon. The Black Sun appears due to the self-projection of the atmosphere onto the sky at the KH point. At the KH point, the Sun disappears from the sky which makes the projection of the atmosphere without the Sun. The absence of the Sun appears in the sky as a black disc. On the Moon there are numerous elements of each type. These elements have different sizes. The elements on the Moon's surface are not always clear, but are more or less altered by their overlap with other elements.

Solar Spider: In the image you can see the Moon in the night sky. Several elements of the same type are circled in this image. We have named this type of element the Sun Spider because it is roughly shaped like a light spider and is a projection of the Sun and its rays. The Sun Spider is made up of the Sun's disc, the cosmic ring that mirrors the cosmic atmosphere and the Sun's rays passing through the Earth's atmosphere.

X. The Moon

In the following picture we have outlined the solar disc and the outer edge of the cosmic ring to make those two elements more distinguishable.

There are a lot of solar spiders on the Moon's surface. Many of them are very small and so they are hard to spot. Even the ellipsoidal features at the bottom edge of the Moon are sun spiders. They are very small and because they are projected onto the curved surface of the Moon, they appear to be ellipsoidal. To understand experimentally that a circle projected onto a curved surface becomes an ellipse you can do the following experiment. Draw a circle on a piece of paper! Then fold the paper into a cylinder so that you can see what you have drawn on the paper. When you look at what you have drawn on that paper you will see an ellipse instead of a circle.

X. The Moon

Black Disk: Images 1 and 2 show the Moon in the night sky. There are darker areas on the Moon's surface than on the rest of the surface. Some of these dark areas are shaped like a circle or an ellipse which is very close to the shape of a circle. In image 2 we have highlighted some of the dark areas so that they can be seen better. These areas are projections of the Black Sun. The Black Sun is a black disc. It appears on the Moon sometimes in the shape of a circle.

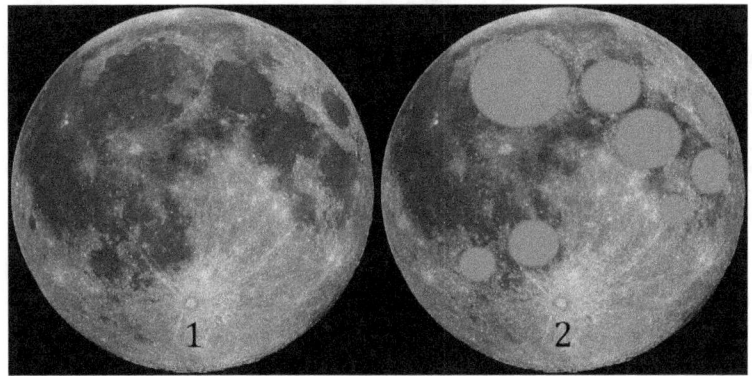

When projected onto the curved surface of the Moon, it appears as an ellipse because a circle projected onto a curved surface becomes an ellipse. Some projections of the Black Sun have their edges altered by other elements and so the black disc is no longer perfectly round. There are many projections of the Black Sun on the Moon.

The Moon's motion in the sky: The solar and lunar disks are parallel.

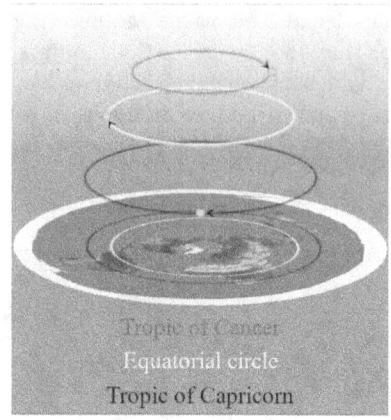

The solar disc is above the lunar disc. As a result, the Sun and Moon perform similar motions in the sky. In this chapter I will explain the motion of the Sun and the Moon to understand the similarities and differences between the two motions. I will start by describing the movement of the Sun in the sky. In the image, the Sun's motion is shown without considering two principles of perspective. These two principles are as follows: a. the shrinking of the trajectories with increasing altitude, and b. the curvature of the atmosphere and its transformation into an apparent dome. The sun spirals clockwise up and down above the flat planet on the surface of an imaginary truncated cone with the bottom base above the Tropic of Capricorn and the top base above the Tropic of Cancer. The radius of the Tropic of Capricorn is greater than the radius of the Tropic of Cancer. The orbital-tangential velocity depends on the radius. Consequently, the Sun's orbital-tangential velocity becomes higher and higher as the altitude at which the Sun is located decreases.

The orbital-angular velocity remains unchanged all the time. The Sun's trajectory is a circle. The circle is 360 degrees. The Sun makes one complete rotation in 24 hours. The Sun's orbital-angular velocity

X. The Moon

is found by dividing 360 degrees by 24 hours. Therefore, the Sun's orbital-angular velocity is 15 degrees per hour. Expressed in radians, the Sun travels 2π in 24 hours. Therefore, the Sun's orbital-angular velocity expressed in radians is $2\pi/24$ hours = 0.26 radians per hour. Since the Moon is below the Sun, it also moves simultaneously with the Sun on an imaginary truncated cone which is similar to the imaginary truncated cone on which the Sun moves, but translated downwards. In this case, the Moon and Sun have the same orbital-tangential velocity and the same orbital-angular velocity.

In the following, I will describe the Sun's motion taking into account principle a and disregarding principle b. In this case, the Sun's trajectories get smaller and smaller compared to its real trajectories as its altitude increases. The Sun spirals clockwise up and down above the flat planet on the surface of an imaginary truncated cone with the high base above the Tropic of Capricorn and the high base above the Tropic of Cancer. The radius of the orbit in which the Sun moves gets smaller and smaller compared to the actual radius of the same orbit as the Sun's altitude increases.

Since the radius of the orbit changes with the Sun's altitude, the Sun's orbital-tangential orbital velocity also changes, but the Sun's orbital-angular velocity remains constant. As the Sun climbs the truncated cone, the Sun's orbital-tangential velocity appears to become less and less relative to its true orbital-tangential velocity. When the Sun descends the truncated cone, the Sun's orbital-tangential velocity appears to become increasingly faster than its previous orbital-tangential velocity. From this we can see that the perspective changes the orbital-tangential velocity relative to the actual orbital-tangential velocity. However, the change in this velocity is only an appearance produced by the perspective. As mentioned, the Moon is actually under the Sun. Consequently, the Moon also spirals clockwise up and down above the flat planet on the surface of an imaginary truncated cone with the high base above the Tropic of Capricorn and the high base above the Tropic of

Cancer. The axis of the imaginary truncated cone on which the Moon moves is the same as the axis of the imaginary truncated cone on which the Sun moves. Because the Moon is below the Sun, the imaginary truncated cone on which the Moon moves is at a lower altitude than the imaginary truncated cone on which the Sun moves.

The radius of the orbit in which the Moon moves gets smaller as the altitude of the Moon increases. Since the radius of the orbit changes with the altitude at which the Moon is located, the orbital-tangential velocity of the Moon also changes, but the orbital-angular velocity of the Moon remains constant. As the Moon climbs the truncated cone, the Moon's orbital-tangential velocity becomes slower and slower, and as the Moon descends the truncated cone, the Moon's orbital-tangential velocity becomes faster and faster. From this we can see that perspective changes the Moon's orbital-tangential velocity. But the change of this velocity is only an appearance produced by the perspective.

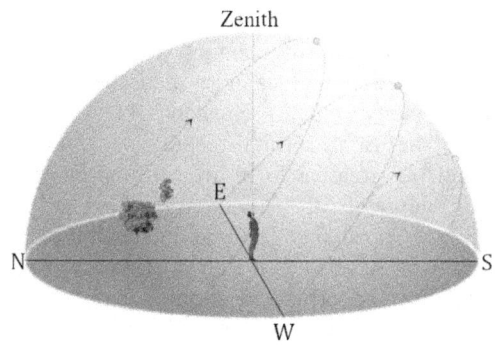

In the image above, the Sun's movement is represented taking into account both of the above-mentioned principles of perspective. The ellipse representing the flat earth does not represent the whole planet, but only the portion of the earth having a radius equal to the distance

X. The Moon

from the observer to the horizon line. The portion of the earth is shown to be flat, so it is represented as it really is. The perspective, however, curves it so that it becomes an inverted dome, so it becomes a concavity. The inverted dome appearance of the earth's surface can also be seen in the middle of the ocean and from an aerial perspective, for example from the perspective of a man in a hot air balloon at an altitude of about 10 km. In the image, the atmosphere is curved by perspective and thus becomes an apparent dome.

The height of the apparent dome varies according to the altitude at which the observer is standing. The maximum height of the dome on which the Sun moves is equal to the maximum altitude of the Sun. The Sun is at maximum altitude when it is at aphelion. The minimum height of the dome on which the Sun moves is equal to the minimum altitude of the Sun. The Sun is at minimum altitude when it is at perihelion. The same applies to the Moon.

The maximum altitude of the dome on which the Moon moves is equal to the maximum altitude of the Moon. The Moon is at maximum altitude when it is at apogee. The minimum height of the dome on which the Moon moves is equal to the minimum altitude of the Moon. The Moon is at minimum altitude when it is at perigee. The height of the dome on which the Moon moves is lower because the Moon is always below the Sun.

The sun rotates on the apparent dome as in the picture. The trajectories in the image are the paths of the Sun as seen from locations on the northern hemiplane. The trajectory on the left is the path of the Sun during the solstice on 20 June and 21 June respectively. During this period, the Sun rises in the northeast and sets in the northwest. The middle trajectory represents the path of the Sun during the equinoxes on 20 March and 22 September and 23 September respectively. During this period, the Sun rises in the east and sets in the west. The trajectory on the right is the path of the Sun during the solstice on 21 December and 22 December respectively.

During this period, the Sun rises in the south-east and sets in the south-west. There are three points to note here:

1. the visual sense cannot distinguish the growth of the apparent dome. In other words, to the visual sense the apparent dome is always the same height.
2. because the height of the apparent dome appears to be the same, the radius of the orbit in which the Sun moves appears to be the same.
3. On the northern hemiplane, during the June solstice, there is sunlight for a longer period of time than during the December solstice. In other words, the Sun stays in the sky longer during the June solstice than during the December solstice.

These three aspects give the impression that the Sun would change its orbital-angular velocity so that during the June solstice it would be slower than during the December solstice. Exactly the same is true of the Moon. For the visual sense, the Moon also changes its orbital-angular speed. Because of the difference in altitude between the Sun and the Moon, the Sun's orbital velocity appears to be different from that of the Moon. If the orbital velocities of the two entities differ, then the two entities may have different positions in the sky. This is why the Moon and Sun may have different positions in the sky.

The following images show both the apparent dome of the Sun and that of the Moon. In the first image, the Moon is partly in the truncated cone of sunlight. Because the truncated cone of sunlight is brighter than the light of the Moon, the part of the Moon's disc that is in the truncated cone of sunlight can no longer be seen. In other words, the Moon's disc can only be partially seen.

X. The Moon

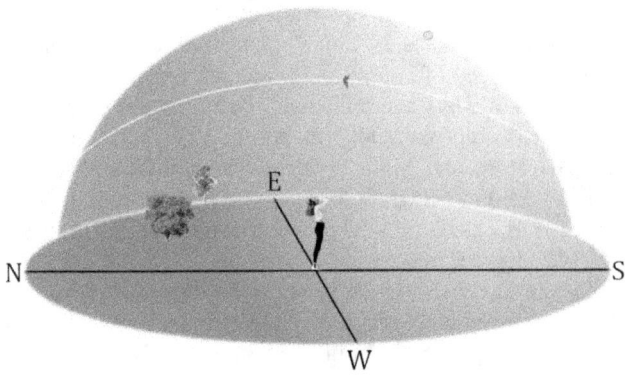

In the next image, the Moon has fully entered the truncated cone of sunlight and overlapped the solar disc. The light from truncated cone of sunlight is brighter than the moonlight which makes the Moon appear to disappear.

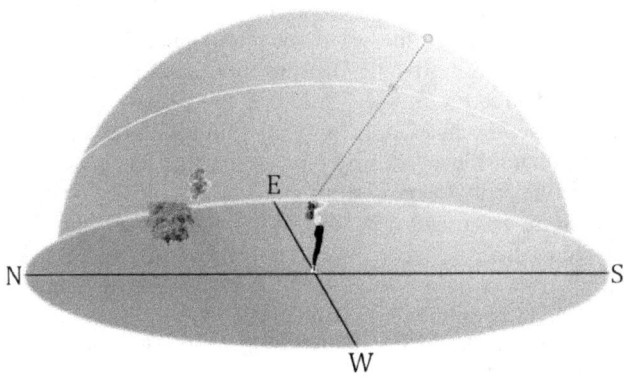

The trajectory of the Sun is always a circle. A circle is always 360 degrees. A solar day lasts 24 hours. Therefore, the Sun rotates at an orbital-angular speed of 360/24 = 15 degrees per hour. The trajectory of the Moon is always a circle. A circle is always 360 degrees. A lunar day lasts approximately 24.83 hours = 24 hours and 50 minutes. Therefore, the Moon rotates at an orbital-angular speed of 360/24.83 = 14.4966 degrees per hour. We will round 14.4966 to 14.5 for ease of calculation. The Moon rotates at an orbital-angular speed of 14.5 degrees per hour.

So the Sun's orbital-angular speed is 15 degrees per hour and the Moon's is 14.5 degrees per hour. The Moon loses a distance in degrees of 0.5 degrees per hour from the Sun. The Moon loses a distance in degrees of 0.5 x 24 = 12 degrees per day. A calendar month has about 30 days. 12 (degrees) x 30 (days) = 360 degrees. This means that in 30 days the Moon is back where it started. This distance in degrees, equivalent to 12 degrees per day, that the Moon loses from the Sun causes the phases of the Moon. The Sun and Moon rotate clockwise.

The Sun casts a truncated cone of light onto the surface of the flat planet, which I will call the solar truncated cone. Because the Moon lags behind the Sun by 12 degrees per day, it moves counter-clockwise towards the Sun. By the time the Moon reaches the solar truncated cone, because it is only a dimmer light than the solar truncated cone, it diminishes by the portion that has passed the solar truncated cone. The Moon is getting closer and closer to the Sun and so losing more and more of its disc. When the Moon is superimposed on the solar disc, it becomes invisible because sunlight is much brighter than moonlight. When the Moon overlaps the solar disc, the Moon is said to be in the new moon phase.

Phases of the Moon: In simplistic terms, the phases of the moon are as follows: full moon, waning moon, new moon and waxing moon. The full moon occurs when it is in a position opposite the sun. In our case, the moon in position 1 is the full moon. After the full

X. The Moon

moon phase, the moon decreases continuously until it overlaps the sun. Therefore, the moon in positions 2, 3, 4 and 5 is called the waning moon. At position 6, the moon overlaps the sun. Because the sunlight is much brighter than the moonlight, the moon can no longer be seen and seems to disappear. This is the phase of the moon called the new moon. After the new moon phase, the moon grows steadily until the full moon phase.

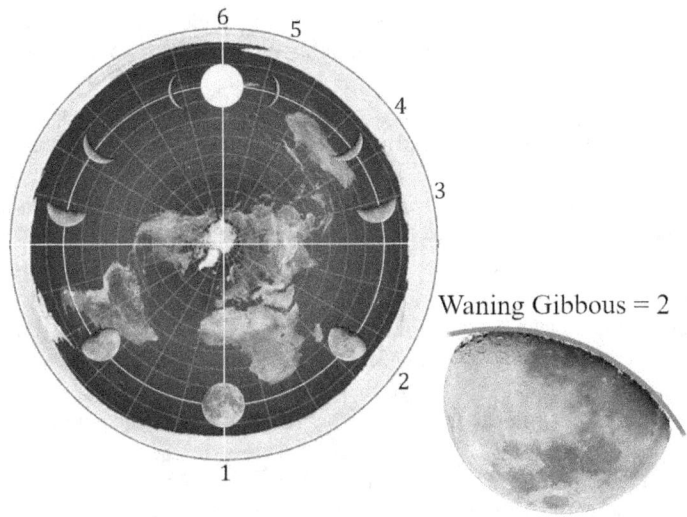

Waning Gibbous = 2

During the time between the new moon and the full moon, the moon is said to be in the waxing phase and is called the waxing moon. The picture above shows the relative motion of the Moon lagging behind the motion of the Sun. In other words, we can see the Moon's relative lagging motion in the following way: while we imagine the Sun to be stationary, the Moon is moving counter-clockwise due to its lagging motion relative to the Sun's motion. In the first phase, expressed by the 6th position of the Sun and the 1st position of the Moon, the

Moon is in the spot with the greatest degree of darkness. The Moon therefore appears in its entirety. In the second phase, expressed by the 6th position of the Sun and the 2nd position of the Moon, the Moon enters the solar truncated con and loses some of its light. This phase of the Moon can be called Waning Gibbous because the Moon is in its waning phase and the edge of the Moon, resulting from the intersection of the Moon with the solar truncated cone, is gibbous. The edge of the Moon resulting from the intersection of the Moon with the solar truncated cone is the convex edge. The Moon appears as a circle in the sky and the horizontal section of the solar truncated cone is also a circle.

By intersecting the Moon, thus a circle, with the solar truncated cone, also a circle, we would expect the convex edge to be concave rather than convex. The circle of the Sun's truncated cone intersecting the Moon's circle is very large and therefore the line of intersection of the two circles is approximately a straight line. This straight line is curved by the retina. Therefore, the line of intersection between the circle of the solar truncated con and the circle of the Moon appears as a convexity. In the third phase, expressed by the 6th position of the Sun and the 3rd position of the Moon, the Moon enters further into the solar truncated con and loses more of its light. In the fourth phase, expressed by the 6th position of the Sun and the 4th position of the Moon, the Moon enters further into the solar truncated cone and loses more of its light. In this phase, the concavity of the edge resulting from the intersection of the truncated con of the Sun and Moon is clearly visible. In the fifth phase the same phenomenon occurs, i.e. the Moon decreases and the concavity of the Moon becomes greater than the previous one. In the sixth phase, expressed by the 6th position of the Sun and the 6th position of the Moon, the Moon overlaps the Sun. Because the Moon's light is much less bright than the Sun's, the Moon seems to disappear.

XI. The Black Sun

In order to explain the Black Sun I think it's good to return to the structure of perspective.

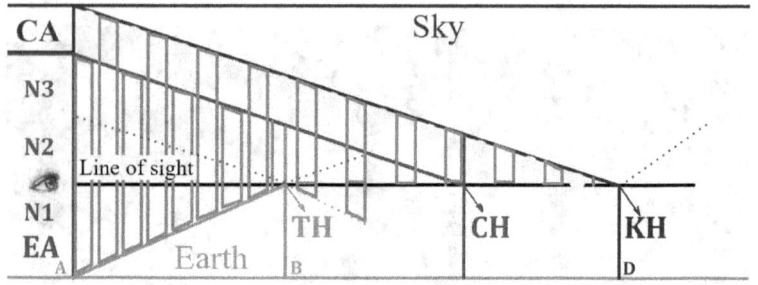

CA = Cosmic atmosphere
EA = Earth atmosphere

TH = The terrestrial horizon
CH = The celestial horizon
KH = The cosmic horizon

On the portion beyond the Cosmic Horizon (KH), the cosmic atmosphere is projected onto the sky. On this portion, the solar disk is below the horizon line and below the line of sight. It is no longer in the sky. When the cosmic atmosphere is projected onto the sky, because the solar disc is missing, a black disc will be projected instead of the solar disc. This black disc produces solar and lunar eclipses. The ancient name for the black disc is the Black Sun.

XII. Mysteries of the Stars

What do we see if we look above the horizon? According to scholastic astronomy, we see isolated entities such as the Sun, the Moon, planets and stars. What the scholastic astronomers call planets and stars, are, however, quite different from what these so-called astronomers tell us. In order to understand this very well it is good to know some very preliminary information. Below I will give you this information and then reveal the mystery behind these entities that people call planets and stars.

Illuminated entities: Above the horizon we can only see bright or illuminated entities whether it is day or night. We cannot see unlit entities, even if they are in the sky during the day. Unlit entities cannot be seen in the sky because of the dense atmospheric layers between the observer and these entities. These atmospheric layers completely cover the unlit entities.

The atmospheric layers, being very dense, cover the entity, making it disappear from our eyes. To better understand this phenomenon I will use an example from painting. Imagine painting a blue sky on a canvas! After you have finished painting the sky, you paint a black disc. To simulate the atmospheric layers that exist between you and the disc you paint a blue layer over the black disc with a brush. After this layer dries, you apply another blue layer in the same way and so on. At some point, the black disc will disappear under the blue layers and all you will see is the sky on the canvas. The same happens in reality. The atmospheric layers that exist between you and a celestial entity do not allow for visual perception of that entity if that celestial entity is not bright or illuminated. Remember this important point: only bright or illuminated entities can be seen in the sky, whether it is day or night.

Circular entities: All entities in the sky appear as smaller or larger circles. Some entities, such as stars, appear as bright points, but the point is also a circle, even if it is a very small circle. Larger circles are called planets, smaller circles are called stars.

Consequently, all entities in the sky appear to be circles that are isolated from each other by very large black spaces.

Visual sense: A brief understanding of how visual perception works is also important in this context. Humans visually perceive shapes around them through their eyes and brains. Light reflected from the visualised shape enters through the cornea, then the pupil and is projected onto the retina where it forms an inverted image of that shape. This image is converted into electrical signals and sent to the brain for interpretation. Some images are easy to interpret, others are very difficult to interpret. For example, the shape of a tree is very easy for any human to interpret. Any human can immediately understand what they are looking at when they see a tree. A star, however, can be very difficult to interpret. The average person thinks that the star is a plasma sun that is spherical in shape, not because he has perceived and understood it himself, but because someone else has told him so. But the star is by no means a plasma sun, it is something else entirely, as we shall see.

Retina: The retina is a layer with a thickness of maximum 0.5 mm. This layer is made up of photosensitive cells and is located on the inner side of the eyeball wall. In adult humans, about 72% of the retina is shaped like a sphere. This approximate sphere is about 22

XII. Mysteries of the Stars

mm in diameter. Because the retina is approximately spherical and the celestial panorama is projected onto the retina, this panorama is perceived as spherical. Imagine you are in a tunnel whose entrance is a square opening! The further you move away from this entrance, the rounder the shape of the entrance tends to become. Another example! Imagine you are looking at a cogwheel that is moving further and further away from you! The teeth of the wheel will get smaller and smaller until they disappear making the cogwheel appear to be round and toothless.

The plane's trajectory: Even the rectilinear trajectories of airplanes are transformed by the retina into round trajectories. Imagine following the trajectory of a plane flying at the same altitude, in a straight line and passing overhead. When the plane appears on the horizon, it appears to be moving in an upward circular path until it reaches a maximum point, after which it appears to be moving in a downward circular path, even though the plane is always flying in a straight line at the same altitude and parallel to the ground. The sun and moon move in the same way in the sky. The Sun, for example, appears on the horizon and then rises higher and higher in the sky following an arc-shaped trajectory until it reaches above the observer's head. It then descends, completing the arc of the circle described on its ascent until it reaches the horizon. Note: entities at very great distances are always transformed by the human eye into circular entities.

Types of light: There are two types of light: emitted light and reflected light. Emitted light is, for example, sunlight, the light emitted by a torch or the light emitted by a candle flame. Reflected light is light that comes from the surface of a body illuminated by a light source. Through an emitted light you cannot see, no matter how dim that light is. You can't even look through a candle flame. Light reflected from a surface reveals details of that surface.

Outer space: The convex-heliocentric theory assumes that there is very low density and pressure in outer space so that there is a near perfect vacuum. This vacuum in outer space is supposed to be much closer to a perfect vacuum than any artificially created vacuum. The Karman line, which would be at an altitude of about 100 km measured from sea level, would be the boundary between the Earth's atmosphere and outer space. This line would therefore mark the boundary between the vacuum in outer space and the Earth's atmosphere. As we all know, the vacuum obtained in the laboratory is obtained in a closed space called a vacuum chamber, and this space is perfectly isolated from the outside environment. The Karman line is only an imaginary line and therefore cannot perfectly isolate the supposed vacuum in outer space from the Earth's atmosphere. There is no perfect vacuum in outer space, but a gaseous atmosphere that gets thinner and thinner the higher the altitude it is at. What scholastic astronomers call the Karman line is in fact the maximum altitude that chemical-powered flying machines and balloons filled with carrier gas can reach. The cosmic atmosphere has varying degrees of translucency and permittivity. Some parts of it are opaque, making it impossible to see from Earth because it is neither bright nor lit. Others are more translucent and permissive so that the light and warmth of eternal fire can pass through them.

Nadir and Zenith: The Nadir-Zenith line is a line perpendicular to the Earth. It passes through the vertical axis of the observer. The Nadir is the point of intersection between the observer's vertical and the Earth's line. It lies below the observer. The zenith is the point of intersection between the observer's vertical and the celestial line. It lies above the observer. The nadir and the zenith are diametrically opposite points.

XII. Mysteries of the Stars

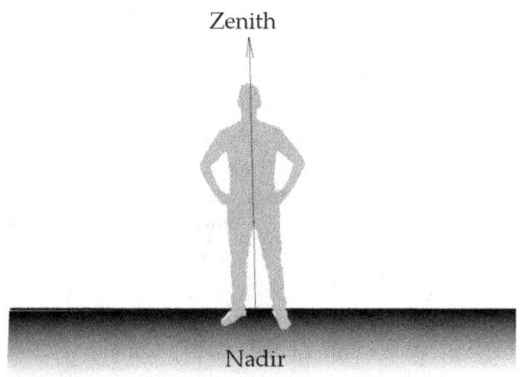

Now we are ready to see what we perceive visually when we are in outer space and looking in the direction of the Earth. Outer space begins at about 100 km altitude. Theoretically speaking, in order to see the planet from outer space, if it were a convex planet with a diameter of 12,742 km, we would need to be at an altitude of about 15,000 km. Let's assume we are at an altitude between 15,000 and 30,000 km. We are in outer space. There is gas in space and it is very dark. The gaseous surface in front of us does not have the same density everywhere. Some areas of the surface are denser, others less dense. Through dense areas of the cosmic atmosphere light penetrates very little or not at all, while through less dense areas light penetrates more or less. Light comes from Earth.

From some areas of the Earth the light will be stronger, from other areas the light will be weaker, and from other areas there will be no light at all. If there is no light coming from the Earth, we will not be able to see anything at all, so we will not be able to see the ground or the water on Earth. If light comes from the Earth we will see larger or smaller areas of light. That's all we will see and nothing more. These areas of light, especially large ones, will have shades of more or less light or more or less shadow. Because the human eye transforms entities at large distances into larger or smaller circles, we will only see larger or smaller circles of light. The circles of light will resemble the planets in the sky. As a result, we will see exactly what we see from Earth when we look at the clear sky at night. The so-called planets in the sky are not solid bodies, but are lights coming from the exosphere of distant territories on Earth. Let's get back to Earth! Let's remember that to the human being on an Infinite Earth, the Earth seems to be tilting continuously towards the horizon. Because the Earth is infinite, it continues beyond the horizon. Consequently, no matter which direction we look at the sky, we are looking at both the sky and the Earth at the same time. If you looked at the sky, if you could remove the Earth's atmosphere, if you could remove the gaseous atmosphere of outer space, and if you had a telescope with which you could see far away, then you could see the far regions of the Earth. The so-called planets in the sky are not spherical, isolated bodies, but are lights in the form of luminous surfaces coming from the exosphere of distant Earth territories. These lights can be seen because they pass through the low-density gaseous spaces of outer space and reach us.

To understand experimentally that planets are distant lights coming from the skies of distant Earth territories, I propose an experiment. Cut a disc out of a piece of cardboard! You can also cut out another shape, but I've chosen a disc to make the explanation easier. At the edge of the disc make a hole about 1-2 cm in diameter! In the centre of the disc stick a needle, a pin or anything else that represents the axis of the disc! Suppose you are in nature in broad daylight and in front of you are: blue skies, a distant forest and a lake, and you are sitting on the sand. Hold the disc in your hand so that the hole is vertical and at the top of the disc, and through the hole you can see the sky. Spin the disc around the axis so that you can see, one by one: the green of the distant forest, the blue of the lake and the colour of the sand. The disc represents the gaseous atmosphere of outer space. The hole in the disc represents a low-

XII. Mysteries of the Stars

density gaseous space in outer space. The lights of the blue sky, the forest, the lake and the sand represent the lights coming from the Earth's distant atmospheres, more precisely they represent the lights coming from the upper part of the Earth's exosphere. Now you can understand very well why the so-called planets change their colours and light intensity. Planets change colour because the low-density gas space on the planets' exosphere circulates above that planet and so different coloured light from that planet penetrates through it. These lights penetrate further until they reach us. So-called planets change their light intensity because light coming from the planet's exosphere sometimes penetrates through spaces with different degrees of translucency. Stars are the suns of distant planets, but these suns are not plasma suns, but are caused by the light coming from the eternal fire. These lights can be seen because they pass through the low-density gaseous spaces of outer space and reach us. Each star is a sun revolving above a planet. Several thousand stars can be seen from each planet. From here you can see that the Earth is infinite.

Why are the stars twinkling? Why do their colours and brightness change? Stars twinkle and change colour and brightness because the atmosphere of outer space rotates, letting lights of different brightnesses pass through low-density gas spaces. Now you can understand very well why stars sometimes suddenly appear and disappear. Stars suddenly appear and disappear because outer space and its atmosphere rotate, and the low-density gaseous spaces of outer space hide behind the high-density spaces of outer space so that the lights can no longer be seen. On the northern hemiplane, the centre of rotation of the stars is on the Nadir-Zenith axis and the stars rotate counter-clockwise. On the southern hemiplane, the centre of rotation of the stars is close to the horizon and the stars rotate clockwise. On both hemiplanes, the paths of the stars in the sky are circular. Why does this happen? Remember! We are on the Flat and Immovable Earth! First of all, it should be noted that the paths of the stars are actually circular.

We can't see the actual trajectory of the stars because the planes on which they rotate are not perpendicular to the Nadir-Zenith line, but have different inclinations. Consequently, the trajectories we see should be elliptical. To quickly understand this, imagine you are drawing, or can even draw, a circle on the ceiling. If you are in the centre of the circle, then you will see the circle as a circle. More precisely, if the centre of the circle is on the Nadir-Zenith line, you see the circle as a circle. If you distance yourself from the circle and look at it again, you will see it as an ellipse. In other words, if the centre of the circle is not on the Nadir-Zenith axis, then the circle on the ceiling will appear to be an ellipse. Consequently, we should see the paths of the stars in the sky as ellipses. We, however, do not see the paths of the stars in the sky as ellipses, but as circles because the eye turns these ellipses into circles. We remember that an ellipse seen from a great distance looks like a circle. If you look closely at the paths of the stars, you will notice that they do not follow a completely circular trajectory, so they do not make a 360-degree arc, but more or less a quarter circle, more or less a half circle, more or less three quarters of a circle. Once they have completed their trajectory, the stars disappear. This phenomenon occurs because the stars are spinning through atmospheric zones of outer space that have different densities. When the atmospheric density of outer space allows the star's light to pass through, we can see that star. When the atmospheric density of outer space is so high that it does not allow the star's light to pass through, we can no longer see that star. The star is spinning in two zones of different densities, so we can only see it when the atmospheric density of outer space is low. This is why stars don't always seem to follow a complete circular path.

Why do the stars rotate counterclockwise on the northern hemiplane and clockwise on the southern hemiplane? The plane of rotation of stars is not always parallel to the Earth. It is tilted, in most cases meaning you can see both sides of it. When you are on the northern hemiplane, you look at the top of the plane on which the

XII. Mysteries of the Stars

star is rotating and see that it is rotating counter-clockwise. When you are on the southern hemiplane, you look at the bottom of the plane on which the star rotates and see that the star is spinning clockwise. To understand this very quickly, you can cut out a circular piece of paper and imagine the star rotating around the edges of the paper. Draw a circular arc with a counter-clockwise arrow on the top of the paper. On the bottom of the paper draw an arc with an arrow in the direction of the star's rotation. You will notice that on the bottom side of the paper, the direction of rotation of the star will be clockwise, so it will be the opposite of the one on the top side of the paper.

Why does the centre on the northern hemiplane around which the stars move and which lies on the Nadir-Zenith line move down towards the horizon line along with the paths of the stars when we move to the southern hemiplane and look north? Because of perspective. To quickly understand this phenomenon, look at a street with lanterns on its edges. If you stand in the middle of such a street and look into the distance, you will notice that the lanterns seem to descend towards the ground the further they are from you. The lanterns near you are above your head and the ones in the distance seem to get closer to the ground the further they are from you.

If you are exactly on the North Pole you will see above you the centre of rotation of the stars, which is the North Star, along with the circular paths of the stars. As you move towards the southern hemiplane, you move away from the North Star. That's why both the North Star and the stars appear to move towards the horizon like distant street lamps. The further you move away from the North Pole, the lower the North Star appears in the sky until you reach 20 degrees latitude, at which point the North Star drops below the horizon so that it can no longer be seen.

Why does the Polaris star seem not to move? The North Star is also called the Polaris star. All the stars move because the sky rotates. The higher the stars are in the sky, the smaller their trajectory appears to be. Polaris is the most distant star that can be seen in the sky. That's why its trajectory becomes so small that it appears to be a point. This makes Polaris appear to be motionless.

XII. Mysteries of the Stars

I have a Nikon COOLPIX P900 digital camera. This camera has an optical zoom of 83x and a digital zoom of 332x. I have looked at stars with this camera and noticed that they appear as flickering lights that have a surface on which there are ripples as if looking through water. Isn't there a dome above the flat planet and above that dome there is water? Before answering this question I offer some background information.

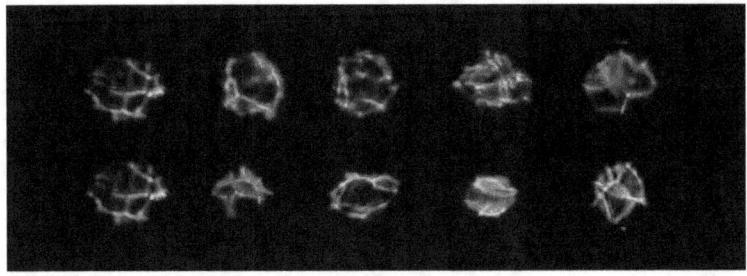

Fluid is a substance that continuously deforms when an external force is applied to it. In the colloquial sense, the term fluid is usually used in the sense of liquid. In the scientific sense, however, the term fluid includes both liquids and gases. Both liquids and gases are fluids because they behave in a similar way. However, there are some differences between liquids and gases. Liquids can form free surfaces, but gases cannot form free surfaces. An example of a free surface formed by a liquid is the surface of a lake. If you look at stars through a high-performance camera, you will notice an effect where stars appear as flickering lights with a surface on which there appear to be ripples. They look like entities seen through water.

Between you and that star there is a very long distance. Along this distance there are many atmospheric layers made of gases. Because gases are fluid like water, they have the same effect as water. The difference between gases and water in this case is that gases are less dense than water and so for them to produce the same effect as water there needs to be a very large distance between you and the object being observed. In previous chapters, we explained that the sky appears to be a dome because the line of sight around you becomes a circular line with a smaller and smaller diameter as your altitude increases. This gives the impression that the sky is a dome. The sky is not a dome. It is made up of the Earth's atmosphere, the celestial atmosphere and the cosmic atmosphere, and it is above a flat, immobile and infinite Earth.

Consequently, there is no dome above our planet and there is no water above this supposed dome. Stars appear as a flickering light that has a surface on which there are ripples because the atmosphere between you and that star is made up of numerous gaseous atmospheric layers that behave like a fluid such as water.

XIII. Questions and Answers

How can it be that when the Sun is at aphelion, i.e. when it is at its highest altitude, it is summer, and when it is at perihelion, i.e. when it is at its lowest altitude, it is winter?

Before answering this question, it is necessary to give you some preliminary information. First of all, it is good to know that Europe is on the northern hemiplane. Secondly, it is good to know what aphelion and perihelion mean. The point at which the Sun is at its highest altitude is called the aphelion. The point of lowest altitude at which the Sun can be is called the perihelion. We have stated that the Sun is moving back and forth on an imaginary truncated cone. This motion of the Sun can be broken down into two subcomponents. The first subcomponent is the Sun's vertical back-and-forth motion. The second subcomponent is an approximately circular motion of the Sun. In what follows I will explain the first subcomponent.

The Sun's vertical back-and-forth motion is due to the cosmic atmosphere, made up of gaseous clouds, performing a vertical back-and-forth motion. The cosmic atmosphere's vertical back-and-forth motion is due to hydrogen coming from the Earth. The hydrogen rises until it reaches the bottom layers of the cosmic atmosphere. Hydrogen accumulates below the cosmic layers as a layer of hydrogen that pushes up the cosmic atmosphere. During this process, the hydrogen layer is growing because hydrogen is continually rising from the Earth. The hydrogen layer gets bigger and bigger and therefore has a greater and greater pushing force. This causes the cosmic atmosphere to move vertically upwards. At some point, the cosmic atmosphere is comasted and therefore can no longer be pushed by the hydrogen layer. The hydrogen layer exerts increasing pressure on the cosmic atmosphere. Because the hydrogen layer can no longer push up the cosmic atmosphere, more and more hydrogen is pushing through the small empty spaces of the cosmic atmosphere. The hydrogen layer is thinning and exerting less and less pressure on the cosmic atmosphere. Because of this, the cosmic atmosphere begins to descend. This causes the cosmic atmosphere to move back and forth vertically.

To experimentally understand the vertical back-and-forth motion of the cosmic atmosphere you can do the following experiment. Make small holes in a balloon and then fill the balloon with hydrogen! Tie the mouth of the balloon and let the balloon fly! The balloon will fly. During the flight, the balloon will lose more and more hydrogen through the small holes. When there is so little hydrogen in the balloon that it can no longer lift the balloon, the balloon will stop flying and fall to the ground. After you recover the balloon you can put hydrogen back into the balloon so that it can fly again. In this way, you notice that the balloon performs a vertical back-and-forth motion. This is the principle behind the vertical back-and-forth motion of the cosmic atmosphere. Since the cosmic atmosphere contains the solar structure and the cosmic atmosphere performs a vertical back-and-forth motion, the solar structure will also perform that vertical back-and-forth motion.

In the following I will explain the second subcomponent of the Sun's motion, namely the rotational motion. The Sun's rotational motion is due to the convection phenomenon produced by solar heat. The Sun heats gases in the cosmic atmosphere and gases in the Earth's atmosphere. Heated gases are lighter than other gases. That's why these gases start to rise. This phenomenon is called convection. Convection causes the cosmic atmosphere to rotate. To understand this phenomenon better, I will give an example. In the following picture there is a candle and a paper spiral. The paper spiral is suspended so that it hangs above the candle. In this case, the candle acts as a heat source. The candle heats the air above it. Heated air is lighter than unheated air and therefore rises. This phenomenon is called convection. The heated air presses against the bottom of the paper spiral, causing the paper spiral to rotate. As a result, the paper spiral rotates due to convection.

XIII. Questions and Answers

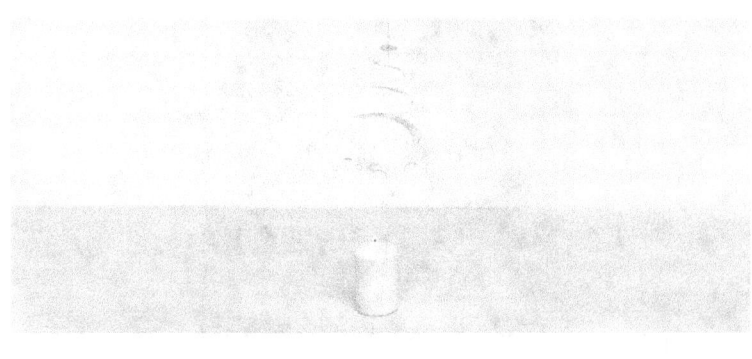

The cosmic atmosphere also rotates due to convection. As the cosmic atmosphere rotates, so does the solar structure. As a result, the solar structure performs two motions simultaneously. The two motions of the solar structure are: roughly circular rotational motion and vertical back-and-forth motion. The rotational motion is approximately circular because the trajectory of the solar structure is not a perfect circle. When the solar structure makes a complete rotation, it will be at a point slightly lower, if it is in the downward phase, or slightly higher, if it is in the upward phase, than the initial point because it is moving in a spiral. Therefore, the Sun's rotational motion is approximately and not perfectly circular. Now it's good to understand a bit about the difference between the Sun's heating power and its illuminating power. The Sun's heating and lighting principle is identical to that of a candle flame. Therefore, to understand the difference between the heating power and the lighting power of the Sun, we can study a candle flame.

Imagine you are in a large room where it is dark! There is no furniture or other objects in the room, only a candle burning. Therefore, the candle flame cannot be blocked by any object. Move as far away from the candle as you can and then look at the candle flame! You will notice that you will be able to see the candle flame, even though you are a long way from it. From here you can see that the illuminating power of the candle flame is very great. Now get close to the candle flame until you feel its warmth with your hands! You will notice that in order to feel the warmth of the flame with your hands, you will need your hands to be about a centimetre away from the candle flame. Consequently, the difference between the lighting power of a flame and its heating power is gigantic.

In other words, the lighting power of a flame is gigantic compared to the heating power of the same flame. In the next few lines I will explain a little about the terms summer and winter. These terms have several definitions. The definition of summer and winter sometimes depends on the country, sometimes on the geographical area and sometimes on the culture of the geographical area. In addition, there are astronomical and meteorological definitions of these terms. In order to be able to explain the occurrence of the seasons on the planet and answer the above question, we need to have a clear definition of summer and winter. I will therefore define

XIII. Questions and Answers

these terms. My definition will only be valid in this context. The definitions of summer and winter are as follows:

winter: winter is the period during which the following calendar months occur: December, January and February.

summer: summer is the period during which the following calendar months occur: June, July and August.

In the following I will explain a little about the shape of the Sun. The Sun is an entity that is shaped like a calotte cut out of a sphere. The shape of the Sun can be broken down into two subcomponents. The two subcomponents are the central area and the peripheral ring. In the following picture you can see the shape of the Sun with its two subcomponents.

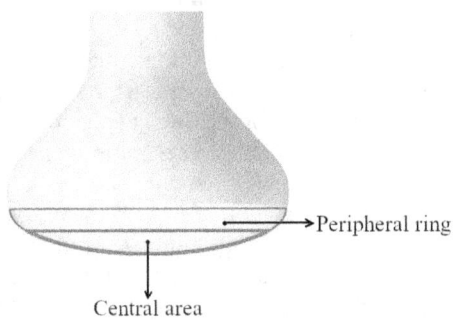

In the following I will look at how the Sun behaves. When the Sun is at aphelion, it is at a point above the Tropic of Cancer. When the Sun is at perihelion, it is on a point above the Tropic of Capricorn. The Sun moves in a clockwise spiral up and down above the flat planet on the surface of an imaginary truncated cone with the top base above the Tropic of Cancer and the top base above the Tropic of Capricorn. The Tropic of Cancer lies on the planet's northern hemiplane. The Tropic of Capricorn lies on the planet's southern hemiplane. The Tropic of Cancer has a smaller radius than the Tropic of Capricorn. The Sun always makes one complete rotational movement in 24 hours. When over the Tropic of Cancer, the Sun is at aphelion. Aphelion is the maximum altitude at which the Sun can reach. When over the Tropic of Capricorn, the Sun is at perihelion. Perihelion is the minimum altitude the Sun can reach. The Sun at aphelion is at the maximum altitude but at the minimum horizontal distance from the northern hemiplane. The Sun at perihelion is at minimum altitude but at maximum horizontal distance from the northern hemiplane. The orbital-tangential velocity of the Sun at aphelion is minimum. The orbital-tangential velocity of the Sun at perihelion is maximum. The Sun at aphelion heats the northern hemiplane with the central zone. The Sun at perihelion heats the northern hemiplane with the peripheral ring. The central area of the Sun emits more heat than its peripheral ring. Summer on the northern hemiplane, and therefore in Europe, is determined by the following factors: the fact that the Sun at aphelion heats the northern hemiplane with its central zone, the fact that the Sun at aphelion has a minimum orbital-tangential velocity which gives it time to heat the northern hemiplane, and the fact that the Sun at aphelion lies horizontally at a minimum distance from the northern hemiplane. Winter on the northern hemiplane, and hence in Europe, is determined by the following factors: the fact that the Sun at perihelion heats the northern hemiplane with its peripheral ring, the fact that the Sun at perihelion has a maximum orbital-tangential velocity so that it does not have the time to heat the northern hemiplane too much, and the fact that the Sun at perihelion is at a maximum horizontal distance from the northern hemiplane. When it is summer on the northern hemiplane, it is winter on the southern hemiplane and vice versa. This interplay between the three factors determines the temperatures on the planet and creates the seasons.

If the Earth were flat, the Sun would have to get smaller and smaller at sunset until it is no longer visible because the distance between it and the observer is increasing. In reality, however, when

XIII. Questions and Answers

the Sun reaches the horizon, it is shaped like a disc and starts to sink. How can this sinking of the Sun be explained? Some believe that if the planet were flat, then the Sun would have to become so small on the horizon that it would have to disappear from view, but could be brought back into view with the telescope. This view is totally wrong. What happens to buildings that sink beyond the horizon also happens to the Sun. The Sun, being a large object in the sky, shrinks to the horizon, but does not become so small that it disappears from the field of view. It passes beyond the horizon as a circle and slowly, slowly begins to sink as the distance between the observer and the Sun increases.

Once the Sun disappears completely below the horizon, this phenomenon is called sunset, it can no longer be brought back into the field of view with a telescope at ground level because, as I said before, sunken objects can no longer be brought back into the field of view with a telescope at ground level, even if that level is at a high altitude such as the crest of a mountain. Objects that have sunk behind the horizon and lie between TH and KH could be brought back into the field of view with the telescope and by increasing the altitude until they return between the observer and TH. The sun begins to sink in the interval between TH and KH. While the Sun is between TH and KH, it can be brought back into the field of view with a special telescope made so as not to damage the eyes, but only from such a high altitude that it returns between the observer and TH. At sunset, the Sun sinks completely behind KH. When the Sun sinks completely behind KH, it cannot be brought back into the field of view in any way.

Why do clouds appear behind the Sun?: There is a view that there could be clouds behind the Sun, i.e. above the Sun if we observe the Sun in its real position and not the one determined by perspective. There are no clouds behind the Sun.

XIII. Questions and Answers

When large clouds pass in front of the Sun, i.e. pass below the Sun when we observe the Sun in its real position and not the one determined by perspective, and these clouds are not very opaque, the portion of the clouds passing in front of the Sun disappears because the strong light of the Sun totally diminishes the colour of these portions of the clouds. The parts of the clouds that pass below the Sun, here I am referring to the Sun seen in perspective, and which are quite opaque remain visible. This effect creates the illusion that clouds are passing behind the Sun, i.e. above the Sun if we look at the Sun in its real position and not the one determined by perspective.

On 14 July 2014, the GoFast rocket was launched. A video shows the rocket colliding with the dome and coming to a sudden stop. How do you explain this collision? There is the idea that the earth would be covered by a dome. The following arguments are usually put forward in favour of this idea: religious arguments, the dome-shaped appearance of the sky, the round shape of the rainbow and the abrupt cessation of flight by the GoFast rocket launched by CSXT (Civilian Space eXploration Team) on 14 July 2014, the abrupt cessation of flight being interpreted as a collision with the dome of the sky.

Note: The picture above shows the working principle of a rocket with chemical propulsion.

The sky appears to be a dome because the panorama of the sky is projected onto the concave side of the retina. According to CSXT, the GoFast rocket would have reached an altitude of 117.6 km. After being launched, it flew up to a certain altitude and then came to an abrupt halt producing the appearance of a collision. Some present the sudden stop of the GoFast rocket as a collision of the rocket with the dome of the sky, and thus as evidence for the assumption that there is a dome on earth. The rocket did not collide with any dome. In the next few lines I'll tell you what caused that apparent collision.

XIII. Questions and Answers

There are two important forces acting on the rocket: the weight of the rocket and the propulsion force. Weight is a force that can be represented as a vector whose direction is vertical and whose sense is downward. The propulsion force is a force that can be represented as a vector whose direction is vertical and whose sense is upwards. In the following I describe how the propulsion force arises.

During launch, the rocket projects a jet of gas towards the ground, creating a force that can be called thrust. The thrust force is a vector whose direction is vertical and whose sense is downward. The earth reacts with a force that can be called the propulsion force. The propulsion force is a force that can be represented as a vector whose direction is vertical and whose direction is upwards. The thrust force and the propulsion force are equal in modulus. The modulus is the numerical value of the vector. The thrust force and the propulsion force do not cancel each other out, although they are equal in mode and have opposite senses, because the thrust force acts on the ground and the propulsion force acts on the rocket. They would only cancel if they both acted on the same entity.

From a certain altitude onwards, the propulsion force is generated according to the same principle as the propulsion force at rocket launch, but this time the gas jet is projected against the air layers. Consequently, this time it is not the ground that reacts with a force called the propulsion force, but the air layers. I remind you that there are two important forces acting on the rocket: the weight of the rocket and the propulsion force. These two forces can be represented by vectors whose directions are vertical but whose senses are opposite. The weight is a downward vector, while the propulsion force is an upward vector. The rocket can only fly as long as the propulsion force is greater than the weight of the rocket.

As altitude increases, the atmosphere becomes increasingly thinner, which means that the gas jet is projected against increasingly thinner layers of air. This causes the layers to react with an increasingly lower propulsion force. In other words, the propulsion force decreases continuously during vertical flight because the gas jet is projected against thinner and thinner layers of air. The GoFast rocket has been launched. Just as the rocket reached an altitude of 117.6 km, the propulsion force was equal to the weight of the rocket, causing the rocket to come to a sudden stop and crash to the ground.

As a result, the rocket did not come to a sudden stop because it collided with a dome, but because the propulsive force was equal to the weight of the rocket. The technology whereby a rocket flies based on the propulsive force resulting from the projection of a jet of gas against layers of air I will call chemical propulsion because it is based on the combination of chemicals.

Based on chemical propulsion, a rocket can only fly up to an altitude of about 120 km. At higher altitudes than this, the air is so thin that the propulsive force becomes less and less until it equals the weight of the rocket, which will cause the rocket to stop and fall. From this we can see that the so-called NASA astronauts couldn't make it past the altitude of about 120 km and therefore didn't make it to the Moon.

Why do the authorities support the convex-heliocentric theory? What advantages do they derive from this fact? Truth raises awareness, while lies lower awareness. A conscious man has no need of authority. Consequently, in order to stay in power, any authority must lie. Lying lowers people's level of consciousness which creates confusion and fog in people's minds so that people can no longer understand reality. Those who can no longer understand reality have only one chance and that is to believe. Belief traps the mind within an ideology. The so-called scientific and religious authorities create the ideologies that people must believe in and offer them to people through educational institutions, books and the media.

XIII. Questions and Answers

Because the authorities created these ideologies, only they are considered competent to explain and interpret them. Many people believe in these ideologies. They are, however, considered incompetent to explain and interpret the ideologies, which means that they have to listen to the authorities. In this way, authority preserves its power. The main advantage of the authority in supporting self-created ideologies is that by doing so they maintain power. The convex-heliocentric theory is also an ideology created and maintained by the authorities. It is in fact a Masonic doctrine. Freemasonry is one of the most influential secret groups. It has infiltrated many branches of our society, including both the astronomical community and space agencies around the world. Believers in the convex-heliocentric theory have the following disadvantages:

They can't develop technologies that involve terrestrial dimensions. You can't build bridges, tunnels or very long railways because, if you take into account the supposed planetary curvature, the constructions won't work, possibly they won't work properly. For example, a railway track that is curved vertically because the construction engineer has introduced the coefficient of curvature of the earth into his calculations cannot be crossed by a train without derailing. If we also take into account the fact that sometimes the rails are curved horizontally, then when the curvature coefficient is introduced, these rails will be curved both horizontally and vertically, which will certainly lead to derailment of all trains using these rails.

They can't sail the oceans and seas, possibly they will have major difficulties when sailing. If you're going to use spherical trigonometry to navigate, you're likely to be shipwrecked or at best find yourself out of location assessment. Many captains who have used spherical trigonometry have deviated massively from their course, and some have been shipwrecked.

They can't develop their long-range aeronautical navigation if they take into account the supposed planetary curvature. For example, if you want to build an autopilot for an airplane and you think the planet is convex, you will have to take into account the assumed planetary curvature and so you will have to program the autopilot to tilt the nose of the plane progressively downwards throughout the flight. If the autopilot will do this, then the plane may collide with mountains which will result in the plane crashing.

They can't build intercontinental missiles if they are going to account for the non-existent curvature of the planet because they will have to build them in such a way that they constantly tilt downwards for the duration of the flight, and if these missiles behave in this way, then they will not reach their target.

Whoever believes that the planet is convex and floating in outer space has no hope of freeing himself from slavery because he has nowhere to go. When humans understand that the Earth is flat, immobile and infinite, they will be able to create high-performance vehicles to travel to other planets which are close to our planet, and thus become free.

The flat planet is larger than the convex planet. Anyone who does not understand this will agree with the propaganda that the planet is overpopulated and must be depopulated by disease, vaccination, wars and euthanasia. The planet is not overpopulated. Besides there are new territories that can be occupied instead of killing billions of people to depopulate this planet.

XIII. Questions and Answers

In addition to the above reasons, there is also a financial one. For example, NASA is an agency of the US federal government and has a budget that comes from citizens' taxes and is about 20 billion dollars a year that increases year by year. By having a space agency, the political power reserves the right to have a budget that can be used at any time in a different way than officially declared. In this way it is enhancing the power it already has.

Imprint

Author: Mihail Ispan
Content and cover design: Copyright © 2022 by Mihail Ispan
Publisher: Mihail Ispan, Georg-Strobel-Str. 38, 90489 Nuremberg, Germany

ISBN: 979-8-3589-5684-1
Imprint: Independently published

This work, including its parts, is protected by copyright. Any exploitation is not permitted without the written consent of the author.

www.ingramcontent.com/pod-product-compliance
Lightning Source LLC
Chambersburg PA
CBHW071403210526
45465CB00001B/230